D0006748

EQUATIONS OF ETERNITY

Speculations on Consciousness, Meaning, and the Mathematical Rules That Orchestrate the Cosmos

David Darling

MJF BOOKS
NEW YORK

Published by MJF Books
Fine Communications
322 Eighth Avenue
New York, NY 10001

Equations of Eternity
LC Control Number 2001098738
ISBN 1-56731-506-2

This edition published by arrangement with Hyperion.

Book design by Richard Oriolo

Manufactured in the United States of America on acid-free paper. ∞

QM 10 9 8 7 6 5 4 3 2

To my parents,
with gratitude and love

Contents

Part III MIND

Acknowledgments

I would like to thank especially my highly talented agent, Patricia Van der Leun, for her tireless efforts on my behalf; my editor at Hyperion, Jenny Cox, for her many astute comments and criticisms; my mathematical friend, Andrew Barker, for his helpful advice with the second section of the book; and, by no means least, my family for their unfailing support.

Equations are more important to me,
because politics is for the present,
but an equation is something
for eternity.

—ALBERT EINSTEIN

Introduction

You are roughly eighteen billion years old and made of matter that has been cycled through the multimillion-degree heat of innumerable giant stars. You are composed of particles that once were scattered across thousands of light-years of interstellar space, particles that were blasted out of exploding suns and that for eons drifted through the cold, starlit vacuum of the Galaxy. You are very much a child of the cosmos.

In giving birth to us, the universe has performed its most astonishing creative act. Out of a hot, dense melee of subatomic particles—which is all that once existed—it has fashioned intelligence and consciousness. Some of those tiny, primordial pinpoints of matter from the infant cosmos have become temporarily arranged to make your brain and mine. Your thoughts at this very moment derive from energy transactions between particles born at the dawn of time. Somehow the anarchy of genesis has given way to exquisite, intricate order, so that now there are portions of the universe that can reflect upon themselves and ask: Why am I here? What is the purpose of life, consciousness, and reality?

In posing these questions, we are, in a sense, the universe questioning itself—a most extraordinary realization. It helps dispel permanently the notion that we are irrelevant and insignificant in nature's broad scheme. The fact is, we stand at the known apex of cosmic evolution. Small though we may be physically, we are giants when measured on the scale of complexity. And it is that complexity, of our brains in particular, that is an essential prerequisite to awareness.

Yet the universe did not set out to be aware. During the first few chaotic microseconds, when all the matter and energy there would ever be was erupting from the primeval fireball, there was no great plan to make conscious minds. Nature is congenitally blind. Evolution is not, and never was, a steady march toward a certain type of order, or life, or consciousness. There is no way of knowing in advance what forms nature will take, no favorites, no movement toward a predetermined goal.

On the other hand, it is hard to believe that we are here by chance. *Why* are we aware? Why has self-consciousness come about? The most banal (but not the least pertinent) reply is that if it had not, the question could never be asked. But that is hardly satisfying. At least in principle, we can envisage all sorts of possible universes in which there were no brains, no intelligence, no life of any kind. Why does the only universe we know happen to be capable of knowing itself?

In contemplating this puzzle, we are inevitably drawn to others. What purpose do brains and awareness serve? In Darwinian terms, they must somehow carry survival value. But what? And how did consciousness ever get off the ground? That last point leads to one of the most difficult and elusive of problems: How does awareness—and, in particular, self-awareness—work? How can the universe simultaneously exist and in some corner of itself (our heads) form a self-reflexive, self-aware model of itself?

We are all involved in an unfolding of mind. Yet it is a process that until the early years of this century appeared largely irrelevant on the cosmic scale—an interesting but ultimately unimportant byplay of matter. It seemed that, despite the sophistication of our brains and of our thoughts, despite every appearance that we possessed free will, we were nevertheless trivial cogs in some vast cosmic machine.

Now we have seen the demise of that sterile, classical universe. It has come at the hands of quantum mechanics, the scientific theory that currently provides our understanding of the subatomic world. Quantum mechanics has proved to be extraordinarily successful on a practical level. It underpins recent developments in computers, lasers, telecommunications, and genetic engineering. Most important, it is in complete accord with every experiment devised to test it. It works—remarkably well. But implicit in the foundations of quantum mechanics are a number of almost unbelievable concepts. One of these is that the observer of a phenomenon is intrinsically bound up with the results of what he sees. Put more bluntly, in the standard view of quantum physics—the so-called Copenhagen interpretation, developed by Niels Bohr in the 1920s—there is no material reality except at the instant of observation. There are no particles and no events when we are not watching; so that we, as intelligent observers, may have a crucial role to play in determining what is real outside ourselves.

Every instant, with every conscious act, it seems we

may participate in structuring and materializing the physical world. Furthermore, from observations of the universe at large, it has become increasingly clear that the cosmos is uncannily well suited to our existence. The slightest change in any basic law or fundamental constant would have precluded the development of life. Even the size of the universe and the amount of matter it contains appear to be finely tuned in order that intelligent life can spring up. We are here by nature's special courtesy, while, at the same time, nature seems to be surprisingly dependent upon us.

The fact that we can analyze and understand the universe is remarkable. The physicist Eugene Wigner has commented on "the unreasonable effectiveness of mathematics" in describing the physical world. And it is true that through the precise symbolic language of mathematics we can probe nature's innermost workings. Most amazingly, we can contrive whole areas of mathematics only to find that these abstract, almost playful, inventions of our mind describe to the last detail some aspect of physical behavior. That is as incredible as painting someone you have never seen and having that person arrive the next day on your doorstep.

We are far from irrelevant, it seems, in the scheme of things. At the very least, we imbue nature, through our minds, with form and meaning. And perhaps we do much more. If through conscious observation we create, and if through contemplation we understand what we observe, then we hold the keys—the equations—to eternity in our minds.

Part I

MAN

You can overcome the problem [of relating
our inner experience to external reality] only
if you accept the premise that in some
sense man is a microcosm of the
universe; therefore what man
is, is a clue to the universe.

—DAVID BOHM

1

Brain

Six billion years ago, the atoms now comprising your brain were not only dissociated, they were scattered far across the light-years of an interstellar cloud. The cloud condensed and spawned stars and their worlds. And so, eventually, the atoms of your brain-to-be found themselves on a newborn planet, third out from a youthful Sun. Earth's atmosphere evolved. The primal ocean evolved. Life itself evolved, becoming increasingly complex. And all of it happened on its own, in the black, blank time before there was intelligence or consciousness. Planets came from dust, life from nonlife, brains from strands of biological wire.

The first nervous systems were just pathways along which signals from receptors on the outside of an animal could be routed to produce some predictable action—a retreat or an advance, depending on whether the signal meant danger or the next meal. There are hosts of creatures alive today whose "brains" are no more than this—not true brains at all, but mere conduits for signal routing. That may not seem much; a worm, for instance, can never appreciate Bach or the blues, but at least it can sense the vibrations from the feet of concertgoers nearby and burrow its way to safety.

With the worm it is all reflex and no reflection. And yet, in time, species did emerge on Earth that could do more than just respond robotically to stimuli. Brains became more complex, more capable, almost as if nature were following some preconceived scheme that would lead inevitably to intelligence.

We know, however, that this is not so. Nature follows no grand design. Individuals and species interact among themselves and with their environment; some live to spawn offspring resembling themselves and some do not. The wild card—the source of biological novelty—is genetic mutation. A mutation arises when the DNA blueprint of life is subtly altered, perhaps by being imperfectly copied or through the impact between a gene and an ultraviolet ray. Most mutations prove to be fatally flawed and vanish from the gene pool almost immediately. On rare occasions, though, a mutation comes along that adds to, or improves, what was already there. It may not initially be a big improvement, but after millions of years it is likely to be compounded into something quite astonishing.

Consider the eye, for instance, and in particular the

lens of the eye. This is a fairly recent evolutionary innovation. It probably made its debut with the common ancestor of present-day birds and crocodilians, a few hundred million years ago. Yet one of the most important substances in the eye lenses of birds and crocodiles—a protein called epsilon-crystallin—has been around very much longer. In fact, epsilon-crystallin is identical with an enzyme (a protein that acts as a catalyst) known as lactate dehydrogenase that is vital to the production of chemical energy in even the most primitive animals.

Somewhere along the evolutionary line, a series of mutations cropped up that caused the gene for making lactate dehydrogenase to produce large quantities of this enzyme in the tissue that would form the eye. And it happened because the chemical properties of the enzyme—its stability and behavior in light, for example—were fortuitously suitable for building a lens. In other words, it was a matter of pure luck. There was no conspiracy on nature's part to manufacture lenses. There was nothing ahead of time to say that such structures had to exist. They were discovered, stumbled upon, by nature randomly working its way through countless mutations, and chancing upon a few that happened to benefit an organism. To be able to see and, moreover, to react to what you see is an obvious advantage in the struggle to stay alive. On top of that, if you can focus on an image and thus gain a clearer view of your enemy or prey, then you improve your survival chances still further. Such a beneficial adaptation will be preserved and refined through natural selection.

And that is exactly how it was with brains. Through chance mutations and selection, the primitive neural wiring systems of some creatures long ago began to acquire little extras—a switchboard, for instance, through which all incoming, sensory data first had to pass. Such a structure offered the possibility of a more measured response; one that depended on the type and strength of the signals being picked up. The result may still have been essentially mindless—an autonomic, "knee-jerk" reaction to events outside. But it was a clear step up the scale of sophistication.

If you are a worm that can respond to vibrations by smartly wriggling underground you stand a good chance of outliving your confederates who remain on the surface to be gobbled up or trodden on. The obvious survival benefit of having even a primitive response reflex is what got the nervous system off the ground. But that is still far from consciousness as you and I know it; the multicolored, sound-filled, wonderfully detailed, emotional, *meaningful* experience that our brains somehow manage continuously to conjure up. Where did all *that* come from?

Imagine that, in front of you, are the brains of a salmon, a snake, a crow, and a dog. Quite obviously, they are different, both in size and structure. At one extreme, the fish's brain resembles a snippet of entrail; in fact, it requires a magnifying lens to be properly seen. At the other extreme, the dog's brain looks like a reduced version of our own. Taking the remaining two specimens to be intermediate stages, we can interpret the progression from fish to dog as an evolutionary sequence—a series of snapshots from the childhood photo album of the brain.

The fish's brain is no more than a minute swelling at the end of the spinal cord. Weighing less than a tenth of an ounce, it consists primarily of hindbrain (or brain stem) and midbrain, with only a tiny forebrain at the anterior. Hindbrain and midbrain together have been dubbed the neural chassis, for they are a support structure. Within them are the controls for blood circulation, respiration, reproductive functions, and self-preservation reflexes. They are the body's autopilot. But they never harbored a conscious thought. They never held a dream, or an insight, or a desire. A creature that has nothing but this neural chassis lacks any semblance of awareness as we know it.

It is the forebrain, or cerebrum, that is the seat of the mind—and all that elusive thing implies. Not least, the forebrain serves as the brain's "projection room," the place where sensory data is transformed and put on display for internal viewing. In our case, we are (or can be) actually aware of someone sitting in the projection room, watching the show. But the fish's forebrain is so tiny that it surely possesses no such feeling of inner presence. There is merely the projection room itself, and a most primitive one at that. It has only the crudest emotional fitments, with no facility for replaying the reality film or visualizing alternatives, no means of looking ahead.

As for the snake, that is perhaps further up the ladder of consciousness, for its forebrain is relatively larger—several times larger than its midbrain. So presumably it can experience a richer internal universe than any fish is able to. But there is nothing about a snake, or any other type of reptile, to suggest a highly active mind at work. A reptile can, and will, sit for hours staring blankly into space. And if nothing stimulates it—a passing dainty, the intensifying heat of a desert sun—it does nothing, absolutely nothing. A snake or a lizard lacks the gumption to get up and go on its own. It may have a dim awareness of the world around it, but nevertheless it is a slave to its environment, a robot without an active mind.

So we move on to consider the brains of the crow and the dog. And with these, at last, we find evidence of some dramatic development in the higher cerebral centers. The forebrain of the bird and of the mammal are large and bulbous, and they dominate those parts concerned with autonomic functions. Interestingly, however, all of the more primitive neural components (the spinal cord, the brain stem, the midbrain, and the various subregions of these) are still present. In a sense, a dog brain (and a human brain) has a fish brain deep inside it. That is to say, an advanced brain is a primitive brain with advanced parts added on. As always, evolution has been conservative, economical, building upon what is proven to be sound rather than redesigning the whole from scratch.

Of course, it is one thing to look at a brain, and quite another to be its owner. What does it feel like to be a crow or a dog (or a dolphin or a newborn child)? We could argue at length about which animals are truly conscious and which are essentially as unaware as a stone. There are some who would claim that every living thing, and indeed every *thing*—including stones and molecules and individual subatomic particles—has some degree of consciousness. Others might more conventionally judge, say, a bacterium and an insect to be totally unconscious, a fish marginally conscious, a dog highly conscious, and an adult human demonstrably self-conscious. The fact is, to be really sure of another's quality of awareness you would have to become *it*. But then "you" would no longer be "you," so, upon returning to our own mind, you would have no memories, no linguistic descriptions, nothing with which to compare this strange, new experience.

Whatever it may be, consciousness is not a thing that appears suddenly. It does not abruptly spring into being with the addition of a few extra brain cells or by climbing a few rungs of the evolutionary ladder. And yet, if there was never an actual moment when consciousness was born, there were clearly times we can point to during which it made dramatic leaps forward.

One such period of spectacular mental progress centered on the early mammals. These were creatures that had been driven to a nocturnal lifestyle, because the competition for food at other times, from the dinosaurs and other reptiles, was too intense. Without the Sun on their backs to warm their blood, to stir their metabolism, these diminutive animals had been compelled to evolve their own body thermostats. They had become warm-blooded.

But that innovation brought with it a new problem. To fuel a built-in temperature-control system, you have to eat often and well. Some of these nascent mammals were no bigger than a mouse, so they could not gorge themselves on a single kill. They lost body-heat fast and burnt calories at a prodigious rate. As a result, they had to feed, or be on the lookout for food, continuously while they were awake. At best, fruit and other plant matter could have served them as hors d'oeuvres. For the main course they had to have a regular intake of animal protein, like insects—and plenty of them. They had to catch these skittery morsels in the dark, so that superb night vision, as well as a keen nose and ears, was essential. And what is more, they had to pursue and hunt their prey, actively, relentlessly. Of all the factors, this was the most decisive, because it demanded the fabrication of a more detailed, dynamic mental model of the creature's surroundings—a more potent projection of reality.

To survive, the first mammals had to develop a curiosity about their environment, such as had never before existed on Earth. They had to peer behind corners, under leaves, everywhere, with intense interest, all their waking time, because if they failed to eat each hour they would probably die. Only the best searchers—those with the most inquisitive brains—would live to reproduce. It meant a complete departure from the old stolid, reptilian approach. And it meant, concomitantly, a massive expansion of neural circuitry in the brain.

Billions of extra neurons were needed to integrate the flood of new visual, auditory, olfactory, and tactile data, to unite these senses as they had never been properly integrated in the reptiles, and to transform the resulting complex inner model smoothly, millisecond by millisecond. On top of this, the creature had to supply its own motivation and urge to seek out food. Furthermore, it had to give its young, back at the nest, unprecedented attention, for they, too, had to be kept warm and continuously well fed. While one of the parents stayed with the young, the other had to seek out the next meal. And that is not something you do without a powerful emotional charge from the brain. So this was a time when a shadowy sense of compassion, concern, and other new emotions first appeared. And it happened, as always, because of natural selection—because it was advantageous to survival. Compassion sprang out of a passionless wasteland. Altruism toward a mate, toward offspring, toward a larger community, grew up in the absence of altruism. And, in time, the universe engineered even love, quite unemotionally, because it proved a most effective survival response.

* * *

Deepening emotions, intensifying curiosity, expanding awareness—all these mental developments took root in the burgeoning mind of the early mammals. But *why* did it happen that way? There are other creatures, such as ants, that actively hunt or search for food in a manner analogous to that of a mammal. They survive comfortably into maturity to replenish their kind. And yet an ant has no forebrain. It barely has a brain at all. So why should mammals have adopted a radically different approach and evolved such a remarkably high level of awareness? We could just as easily visualize a planet on which all the creatures, large and small, thrived without ever entertaining a conscious thought—a planet of utterly mindless automatons.

As it happens, of course, almost all organisms alive today *are* utterly mindless. What could be manifestly more successful than the teeming hordes of viruses, bacteria, and protozoa? Yet they have not a brain cell among them. Even if you allow that every vertebrate and every higher invertebrate (such as the octopus and squid) has some level of consciousness, that is still only a tiny minority of all terrestrial life.

Nature invariably progresses along the line of least resistance—the easiest course available to it at any given time. That may seem like an argument *against* the evolution of consciousness. After all, laying down the neural network of a mammalian forebrain hardly seems straightforward. But the alternative would be far more complex. It would be to fill the mammalian niche with a creature whose every action was *genetically preprogrammed.* That might work with animals as small as ants. But for much larger creatures it would be fantastically difficult. As it turned out, it was logistically simpler for mammals to evolve with better brains—brains that allowed an unprecedented degree of behavioral flexibility and decision making.

The same holds true in contemporary robotics and computer science. It is relatively easy to instruct a computer-controlled robot to negotiate a fixed maze or carry out a repetitive task, step by precise step. This corresponds to the naturally evolved genetic program for behavioral control in an ant. But to enable a robot to function well in an unpredictable, changing environment—say, as a domestic butler—is vastly more difficult. In this case, the approach of extrinsic, "anticipatory" programming soon becomes hopelessly bogged down. There are too many alternatives to take account of in advance, just as there are countless unforeseeable situations in life that a creature as advanced as a mammal must face. The solution, in the case of the robot, is to equip it to react flexibly—to furnish it with a computer that can learn from its mistakes, that can progressively assimilate its environment as it matures, and that can retain the plasticity to deal with new situations. It was precisely this solution that nature turned to in evolving the brain.

Now we can begin to see why awareness grew in the first mammals. To occupy their niche, they had to be able to fashion a detailed and *unified,* multisensorial picture of the world around them. The data flowing in through the senses had to be merged internally, and the only way to do that was for the brain to develop a coding system. This involved tagging the incoming data from each sense organ, so that corresponding items of data (giving, say, the exact distance and direction of an insect in the dark) could be matched up. That is an awesome computational problem.

And it is one that had to be solved over and over again with each passing moment.

Once the mechanism was in place—the association areas of the mammalian brain properly wired up and the senses fully integrated—there was something new on this planet. It was the concept of an *object;* an object in space and time. And though the mental counterparts of objects held in the brains of the dawn mammals must still have been vague and only dimly experienced, yet they represented an enormous advance.

Furthermore, to complement and make full use of the improved internal reality, there had to be an enhanced *participation* in that reality. The animal had to *act* inquisitively in order to find enough food to survive, and the simplest way for that to happen was for it to feel inquisitive. Instead of responding passively to isolated stimuli in the way reptiles do, the early mammals became increasingly involved with the mental objects of their experience. They weighed up alternatives, tried out different possibilities (especially through play when young), experienced an inner drive toward some elusive goal. *They felt conscious.*

Each of us is a microcosm of nature, and in more than one sense. First, the same processes—the same types of particles and the forces that act between them—occur in our bodies as in the universe at large. Second, through our minds we reconstruct and mirror in abstract form that which lies outside ourselves. And third, most intriguingly, our own personal evolution parallels the evolution of all life on Earth.

In our physical and mental development, we go through the same basic stages that life did in evolving from the first unicellular creatures four billion years ago. We too start out as single, microscopic cells within a warm, nutrient-rich ocean. We too progress to primitive, multicellular forms—blastulas—in which, initially, all the cells are identical, but then later begin to show an increasing degree of specialization.

Just as, over the eons, creatures on Earth acquired elementary nervous systems and neural swellings that were the harbingers of the brain, so, as embryos, we undergo similar transformations—though billions of times faster. Heart, fingers, eyes, and other purposeful structures emerge from the amorphous cellular mass that was once all we were, echoing the manner in which these organs appeared progressively in the biological history of our planet. The crucial difference, of course, is that there is no trial and error in the development of a baby. The trials and errors have already been made, over the previous billions of years, so that now our genes carry the entire code for making a stupendously complex being without (usually) any of the mistakes or wrong turns that life, as a whole, made along its evolutionary trek.

Through a series of extraordinary contortions and infoldings, the initially round blastula metamorphoses into an increasingly elaborate shape. Where before all the cells were identical, primitive organs begin to appear—a tiny beating heart, a slender thread of nerves that in time will become the spinal cord, and a minute accretion that is the precursor of the brain. The whole astounding process is orchestrated by the phased release of certain chemicals that control the enfolding, or gastrulation, of the blastula and the initial formation of organs and tissues.

Each stage in the development of the blastula, and

later of the embryo, depends on the previous stage's having been successfully completed. Form builds upon form, process upon process, hierarchically, just as, in evolution at large, new, more advanced species emerge only when the lower niches have been filled.

Specifically, in the case of the brain, the development begins with the more primitive parts of the hindbrain and midbrain and then proceeds to the regions associated with higher thought. That is as true of each emergent individual in the womb today as it was of the neurological development of life in general. We acquire, by stages, a spinal cord, a brain stem, a midbrain, a lizardlike brain, a mammalian brain and, lastly, an advanced hominid brain—layer upon layer, working out.

In the final few months before we leave the womb, our brains accrete at the rate of over 250,000 cells a minute, with most of that growth centered in the cerebrum—the large, highly enfolded structure that surrounds and surmounts the rest of the brain. As a result, our latent, prenatal mental capacity overtakes, in turn, that of all the lower animals, the great apes and the ancestral species of man, until at birth our brain is the most sophisticated of any creature's on Earth. Its cell count is around a billion billion. And although this number slowly but steadily declines throughout life that is really not so significant. What is—what is absolutely crucial—is the *malleability* of the human brain, especially in infancy. It comes illiterate, but ready to learn language; uncultured, but within a few years able to assimilate the labyrinthine ways, customs, and essential knowledge of modern man. It comes unaware, but soon, mysteriously, acquires a penetrating awareness.

* * *

Consciousness. What does that mean? By its very nature, it is the most manifest product of the brain's activity. Yet it is also the hardest thing we know to analyze and understand.

Start to think about thinking, to be conscious about consciousness, and you run the risk of tumbling headlong into a whirlpool of self-reflection. Consciousness cannot be studied in the same way as the things you are conscious of. For as soon as you turn your attention on it, it ceases to be consciousness and becomes just another thought or experience.

And there is another elusive aspect to it, because what it means to be conscious is essentially what it is like being you, here and now. Yet, as soon as you think that, here and now are already gone: you, and the world around you, have changed. Consciousness is like a river: ever-flowing, unbroken, unrepeating. And that makes it hard for science, with its predilection for pattern and regularity, to study. How do you pin down a thing whose very nature is change?

To begin to understand consciousness, we have to treat it, not as an object, but as an emergent, high-level process with its own set of rules and properties. That makes it impossible to explain solely in terms of the brain's physiology. Of course, ultimately it does depend, moment by moment, upon trillions of minuscule chemical reactions and electrical fluctuations taking place inside our heads. But, by the same token, so is a great painting dependent on many individual brushstrokes, or a symphony on many separate notes. However, not even a total knowledge of the basic compositional elements would, by itself, help you appreciate the art of Picasso or Brahms. Nor would knowing all the physical machinations of the brain serve to explain consciousness.

What might help is to think in terms of models. Mental models are the internal representations which we, along with other animals and computers, use to control our behavior. The fact is, you could not function without good abstract models of your own body and the world around it.

For example: you have a model in your head, founded upon experience, of what an orange is like. It includes a generalized mental description—round shape; tough orange skin; soft edible segments inside; a taste somewhere between sweet and sour; seeds, and so on. As well as this, you have a higher-level, or more inclusive model for all types of citrus fruits, and you have higher ones still for all types of fruits and all types of food. You have models for trees and works by Dickens and sandy beaches. And, most obviously, you have an especially complicated model for *yourself*. You have a model for what you are and what it is like to be you.

The central question is, why is this self-model—this "you" inside you—the only one that is conscious? In other words, why is it the only one of your many mental models that can (a) be aware of the other models and (b) feel what it is like to be something? Well, perhaps it is not. Perhaps the real truth of the matter is that your self-model is the only one built largely by language. Your self-model—your "I"—is the only one that can actually talk about itself and analyze the nature of consciousness (as your self-model and mine are doing right now). This makes it rather more obvious why all your other mental models seem to be unconscious. They may be unconscious to "you" but not to themselves!

Each of the millions of mental models inside our minds, then, may possess some degree of consciousness—a feeling of being aware. But that is not to imply that any of these models are highly conscious, each in its own way as aware as "yourself." For instance, the models in the lower levels of perceptual processing almost certainly contain no concept of self or action or of an external world. Thus, any consciousness they have must be correspondingly limited. Only the complex model of "self in the world" sustains full reflexive awareness—consciousness of being conscious. It is this that seems to be "you."

People with so-called multiple personalities apparently host several self-models, each of which becomes active from time to time. That is a dysfunctional condition, of course. But we can see now that it may be no more than an extension or exaggeration of the normal case. If what we have posited above is true, then every human brain builds multiple conscious models, though usually there is only one "you"—your unique, ever-changing mental construct of self.

This idea also sheds light on the phenomenon of selective attention—the fact that when "you" (your self-model) turn your attention to something, that thing seems to enter your awareness. It may be that the mental model of the subject has become temporarily annexed to the model of self. That is to say, the model of what you are thinking about has become a subset or a satellite of the self-model so that, for a while at least, it has access to the self-model's power of language. As a result, the subject of your immediate thoughts can be aware of "you" just as "you" are aware of it. This two-way process in which subject and self each act as mirrors for the other may lie at the root of all self-consciousness, a conclusion in accord with the familiar fact that only when you focus your attention on something are you aware of yourself. As soon as you allow your thoughts to drift, to become unfocused—as in daydreaming or meditation—"you" seem to disappear. The self-model becomes detached, bereft of a companion model in which to be reflected. "I was lost in thought," you might later say.

So we have a reasonable working hypothesis. Consciousness is what it is like to have a mental model. And self-consciousness is what it is like to possess a special mental model of self.

Now we must take our quest of this thing called *self* further. And, in particular, we must seek out its roots, both in the individual and in the evolution of mankind.

2

Connections

Eighteen billion years or so of cosmic evolution are celebrated in each newborn baby. The infant emerges, opens its eyes, and sees the world from which it was formed. But of that world—that bright, noisy chaos—it can make no sense. Its initial impression: of a kaleidoscopic totality devoid of structure or meaning.

Even so, without any conscious prompting, the baby's brain is already at work, forging links, discerning patterns. First, it isolates what is most essential: the mother's face, the mother's voice, the mother's breast. Effortlessly, the pristine brain homes in on them as if it knew they would be there. And, of course, instinctively, through its genetic programming, it does. In computer parlance, that is part of the brain's bootstrap sequence: "After the first gulps of air, bond with the mother."

Survival needs seen to, the brain begins to absorb and correlate other stable aspects of its environment: faces and voices; fingers; other shapes, noises, smells, and touches. It starts to recognize patterns and make associations. And all of this it does automatically. Throughout its long evolution, the brain has adapted to take in and make use of what is most imperative in its surroundings. And since it cannot absorb everything it needs to know at once, it has evolved so as to be able to do it in a systematic, progressive way.

Whatever a newborn child does, the universe at large does also, because a human baby—like everything else—is an intrinsic part of the cosmos. That may seem like a strange claim to make, but it is logically and physically sound. A factory in which cars are made is a car-making factory. A planet on which there is life is a living planet because the life-forms are a part and a product of the world's substance. And, on the grandest of all scales, if there is sentience within the cosmos then the cosmos itself is sentient. So we may reasonably view an infant's dawning awareness on two levels: as a consciousness arising in the individual and, simultaneously, in the universe as a whole.

At first glance, these two aspects of an emergent mind may seem to be only tenuously linked. That is because we think of a child as "beginning" at some definite point: either at the moment of conception or within a short period afterward. But the situation is not actually so clear-cut. Every human being, and every human mind, has roots that extend indefinitely far back through time. The genes that regulate all aspects of our physical development, including the prenatal fabrication of our brains, were in existence long before we or our parents were born. Those genes, in turn, evolved, step by step, from more primitive genetic material that can trace its ancestry back to the first biochemical reactions on Earth. And we do not have to stop there. We can carry the search for the ultimate origin of ourselves back still further—back to the very beginning of the universe. Taking this broader, cosmic perspective, we can appreciate more readily that the consciousness of the individual is inextricably tied to the consciousness of the whole.

As life evolved over billions of years, brains and minds emerged through natural selection. Niches opened up that could be occupied by creatures of growing mental sophistication. But the dawning of consciousness took place only very slowly, through tiny increments. In order to be selected, each elaboration of the brain had to offer a new survival advantage while, at the same time, integrating fully with the existing biological machinery. From the owner's point of view, it was essential that any extension of consciousness be added seamlessly to whatever mental facilities were already present.

Today, we can watch an incredibly condensed version of the growth of awareness on this planet, and in the cosmos, in each developing child. To be ultimately successful—to survive long enough to have the chance to pass on some of its characteristics—human consciousness has evolved so as to gradually unfold on an as-needed basis during the early life of each individual. It starts, immediately postpartum, with a rudimentary awareness and recognition of stimuli that are essential for life. This initial awareness stems from the way the brain is wired at birth, which in turn stems from the genetically controlled way the brain develops inside the womb. No conscious effort is involved in this process, just as there was none in the overall evolution of brains. The procedure is one of building a brain that only when complete and thoroughly exposed to the reality outside will be capable of consciousness. The universe fashions a brain, so to speak, in the dark: just as a crystal forms or a star takes shape without an intelligent or willful plan to guide it.

With the embryo only a twentieth of an inch in length, the nervous system makes its appearance. Initially, it is just a flattened sheet, one cell thick: each cell apparently identical. But as these cells further divide, they start to differentiate and drift to specific sites. Before long, the brain emerges in miniature adult form, layer upon layer of neurons, each layer harboring cells of a distinctive type and shape. Then comes the most crucial phase: the formation of connections between the billions of individual neurons. Prior to this, the brain is no more capable of thought than is a heart or an interstellar cloud.

It bids fair to be the most astonishing physical process in the universe. No sooner does a neuron reach its appointed place than it begins to sprout an axon—a conducting pathway that in time will carry electrical signals away from the neuron body to other nerve cells in the brain. Like an amoeba—strikingly so—the tip of the growing axon makes its way along. Thin fingers called filopodia stretch out tentatively from it, withdraw, then stretch out again in a new direction, as if groping for some elusive prey. And that is exactly what they are doing. The filopodia are seeking "sticky" surfaces to which they can bind and then pull toward, thereby elongating the whole axon. These sticky surfaces are specially shaped molecules—chemical signposts to guide the growing nerve cell.

Different groups of axons must be able to recognize different signposts, or else most of the axons in the nervous system would grow to the same place. To deal with this problem, evolution has sited many different receptor molecules on the surface of nerve cells, each of which will stick to only one specific molecule in its environment. The result is that nerve fibers can be guided to any target in the developing brain as long as they have receptor molecules on their growth tips that match the molecular terrain they have to traverse. Each adheres to only one type of guide and ignores the rest.

Having reached their destinations, the first, pioneering nerve fibers establish a framework—a skeletal nervous system—upon which all subsequent fibers can build. On their surface, different pioneers have different types of molecules. Later-growing axons can recognize these distinct molecular types and so opt to grow along the pioneer that will lead them to the correct target. Moreover, they know when they have arrived because, again, specific receptors in the axons recognize molecules that are found only on the correct target neuron.

All this on its own, however, is not quite enough. An axon must do more than just reach the right region of the brain. It must dock at a highly specific part of the target's structure. Such a process is known as topographic mapping, and nowhere is it more graphically illustrated than with the wiring up of the optic nerve.

The eye works like a camera, casting an image of the outside world onto the retina. The neurons in the retina translate the light falling on them into electrical signals that are conducted down the axons of the optic nerve to the brain, where the axons form connections. In the brain region that the eye feeds, the pattern of connections is meticulously ordered, so that the axons make connections that exactly reproduce the patterns of the neurons that they came from in the retina. As a result, the image of the outside world is precisely mirrored in the brain, in the form of electrical impulses.

For this to be achieved, the retinal cells all have to know exactly where they are within the retina. They need an address. At the same time, the neurons in the optical regions of the brain have to work out a complementary set of addresses. The chief task of the retinal nerve fibers as they grow brainward is to remember where they came from in the retina, find an address in the brain that matches up, and make a connection there.

So, by unfolding stages, the brain organizes and interconnects itself. While still in the womb, it prepares itself for what it will encounter when it comes into the daylight. And though, in general, all human brains wire up in the same way, the details are subtly different, just as the overall form and growth processes of snowflakes are alike—and yet no two snowflakes turn out identical.

At birth, the brain's labyrinthine web of cells is joined. But it is joined only to the extent of being completely receptive. The communications channels are wide open to the senses, the neurons organized to accept and process whatever signals come their way. All the topographic maps are in place: the retinas mapped down to the finest detail within the visual cortex (curiously located at the *back* of the brain); the ears mapped so that the auditory fields of each are represented, point by point, in regions (less surprisingly) low down and to the side of the forebrain; and the sensation of touch mapped in a thin band—the somatosensory cortex—running from side to side across the top and center of the brain. This last mapping ensures that, when a baby is born, the pattern of nerve endings across its body is faithfully reproduced in a transverse line across the crown of its brain: toes in the center, then feet, legs, trunk, arms, hands, fingers, face, lips, tongue, in that order, moving gradually to either side of the cerebrum.

As the head of the child emerges from the mother, the brain inside is the most receptive it will ever be. And yet, simultaneously, it is the least aware. Brain there may be, but as yet there is no mind—only the substrate, the potential for it.

Then the sensory onslaught begins and, in response, the initial, postpartum wiring of neurons takes place. These connections made, the brain is changed; it has traded a little of its pristine flexibility for stored knowledge and understanding. Now, further impressions can be assimilated in the light of this newly won intelligence. The brain absorbs, alters, absorbs, alters, again and again, molding itself according to the patterns of energy it receives. And so, through this feedback loop, the brain becomes ever more adept at discerning and classifying patterns in the complexity it sees.

Already, the individual has recapitulated, while in the womb, the physical evolution of all life on Earth. Now it is racing through the stages by which life evolved mentally. From mindlessness to shadowy awareness to consciousness of the world about it to consciousness of self, it moves.

To begin with, the child learns only to recognize signs—familiar sights and sounds. Its experience is filled with vivid sensory and emotional images. Early on, it has perhaps a raw, intuitive understanding of the world around it. But it certainly has no rational understanding. Judged in terms of adult consciousness, it remains essentially unconscious.

For the first two or two-and-a-half years of life, it is so. The world is inexplicable, and yet, at the same time, it is experienced less and less holistically. The child's image-filled reality becomes steadily populated with the images of discrete objects, with things that seem to have their own bounds. Steadily, the child's repertoire of signs, recognizable objects, and relationships between objects broadens, even though there is still little understanding. And then comes a most crucial development: the acquisition of language. Suddenly, the floodgates of the mind are thrown open and the infant begins to put symbols to the signs and objects and actions it perceives. That is another talent for which the brain is exquisitely well suited—and not by chance. The brain develops in such a way that there are certain centers within the cerebral cortex that play a pivotal role in speech. Indeed, they exist for that very purpose. Our brains grow so that they are primed ready to learn a complex spoken language. And the acquirement of that language is paramount, for without it we cannot aspire to full human consciousness.

A sophisticated language, and the mental activity that accompanies it, is one of man's greatest possessions. Perhaps even more so than toolmaking, it is the distinctive human quality. Other animals may be capable of fashioning (admittedly crude) tools. But only man can consistently pass on the know-how of his existing technology to future generations. In this way, our descendants are able to use and improve upon that which language has made available to them.

3

Ascent

Take a newborn child from 100,000 years ago, raise him normally today, and he might become a jazz pianist, an architect, a Pulitzer Prize winner, or a Nobel laureate. Or she might blaze for herself a distinguished career in anthropology or quantum physics. Because the fact is that anything a contemporary child can do a child 100,000 years ago could do equally well if nurtured in the same way. Our brains have not evolved in all that time.

And yet, of course, over the last 100 millennia, there have been changes—prodigious changes. Not in the raw structure of our bodies or brains, but in the way we use them. We have become extraordinarily creative (and destructive), philosophical and inquisitive. It is as if a switch had been thrown. The brainpower—the mental latency—had been there for at least a quarter of a million years, and maybe much longer. But something happened in the fairly recent past that hurled man onto a new course. And since it was not physical, it must have been psychological—a new approach to thinking about the world.

It began with an abruptness that is barely credible. From the first four million years of hominid evolution all we have inherited is stone tools. We know that our ancestors were transforming physically and in their patterns of behavior. Yet the durable products of their industry seem to be exclusively, monotonously utilitarian—scrapers, cutters, axes of stone. After the appearance of the first tools, for 2.5 million years there was remarkably little material progress. And then, just 35,000 years ago, in Europe, came the sudden change. There was an outpouring of invention and artistic expression on a scale previously unknown. Delicate implements of bone and antler, beads, pendants, and other body ornaments, superb sculptures in ivory, masterful cave paintings, the first artworks of a symbolic nature—it all begins to well up from this crack in time 35,000 years ago.

Yet to find the source of the outpouring we have to go back much further. This cultural eruption had begun brewing in the time of *Homo habilis*, or "handy man"—the pioneer toolmaker, two million years earlier. For it was with handy man that the brain reached the stage at which it could not only build mental models of objects but also manipulate those models effectively. It could hold an internal image of, say, a stone, then project how that stone might appear if flaked here and there. It could combine mental models in ways that had not yet been done with actual objects. For example, the habiline brain could conceive that wood, wattle, and stone might make a shelter. Effectively, it could peer into the future and, moreover, envisage itself as an agent in that future. And that was a massive leap forward.

You can see from fossil brain casts of habilines where the vital growth had taken place. The so-called frontal and temporal lobes of the cerebrum are far more prominent in handy man than they are in earlier hominids. The frontal lobe, we know, plays a major role in anticipating future events. It is the "what if?" compartment of the brain—the visionary, the forward planner. Here is where mental models, including our putative self-model, are called together, altered, moved around like holographic projections, and then combined according to certain known (or conjectured) rules of behavior. As a result, predictions and speculations emerge about what may happen and, crucially, what role the thinker may play in determining the outcome.

That last step reveals a major difference between advanced hominids and other species. Any mammal worth its salt can look at two paths up a mountain and assess which is the better to take (the easier to climb, the more protected, and so on). It may also have the wits to base its choice on past experience. But only a human, or near-human, can see that while one path may be better now, the other may become so with a little work—perhaps by moving some rocks or planting a shady grove of trees. Through the circuitry of our enlarged frontal lobes, we can plan for a future in which we consciously take account of our own actions. It is no coincidence that, in our anthropological classification scheme, the first member of the genus *Homo*—*Homo habilis*—is also the first hominid to show a pronounced bulge in the frontal region. This is the part of the brain, more than any other, that makes us human.

However, the enlarged temporal lobes in handy man are also significant. Among other things, the left temporal lobe (in modern brains) is essential for the memory and recognition of complex sounds. Deprived of this region, you would be aphasic—unable to make sense of spoken words. In particular, there is a patch of the left temporal lobe, known as Wernicke's area, that is apparently dedicated to the comprehension of language. A separate patch on the lower left of the frontal lobe, called Broca's area, serves to formulate sentences. And all of this sets the scene for a nice detective story, because in the fossil skulls of habiline man there is a little dip that matches Wernicke's area and a small groove that resembles Broca's area. So we are led directly to the question: could *Homo habilis* talk?

Well, you need more than a brain for speech. You need a voice box, or larynx, and above it, an air cavity, or pharynx. As air from the lungs is forced over the reedlike vocal cords in the larynx, the air in the pharynx is made to vibrate, thus producing a sound. The range of sounds that can be articulated depends crucially on the volume and shape of the pharynx. And that is one of the chief unknowns: we cannot be sure how humanlike was the pharynx of our ancestors.

In the basic mammalian pattern, the larynx locks onto the nasopharynx—the air space at the back of the nasal cavity—during breathing. This creates a direct airway from the nose to the lungs. As a result, the animal can drink and breathe at the same time, because liquid flows through channels on either side of the larynx and nasopharynx. But this arrangement, with a high larynx, so useful in one regard, severely limits the spectrum of sounds the creature can make.

Interestingly, a human infant up to the age of about eighteen months has precisely this kind of elevated larynx, so that, like any other mammal, the baby can breathe while nursing, though its repertoire of sounds is narrow. Then, at one and a half, the larynx begins to sink down the neck, transforming the topography of the vocal tract. The breathing and swallowing pathways now cross above the larynx, with the consequent drawback that food can block the airway and cause suffocation. Yet the benefit of the change is immense. The pharynx is expanded, the vocal tract enlarged, and the youngster can now emit the panoply of sounds that comprises modern speech.

Unfortunately, the soft tissues of the vocal tract, like those of the brain, do not fossilize. So we are left trying to reconstruct the sound-making apparatus of our extinct relatives from what remains—the jaw and the base of the skull.

In one controversial technique, the degree of bend in the skull base is taken as a guide to the position of the larynx. The procedure starts with a comparative study of modern mammals, ranging from dolphins to apes. From this comes a generalization, that a relatively straight, unflexed skull base points to a larynx high in the neck and therefore to a nonhuman vocal tract. Contrastingly, if the skull base is highly flexed, this signifies a lower larynx and a vocal tract capable of producing more humanlike sounds.

The trouble is, no well-preserved skull bases of handy man have yet been found. We only pick up the trail with his more humanlike descendant and our immediate ancestor—*Homo erectus*, or "upright man." And the results are tantalizing. The base of upright man's skull displays bending and flexing comparable to that of a six-year-old child—not an ape, but a young human child. So, if this method is to be trusted, *erectus* had clearly crossed the threshold leading to true language.

Had handy man passed over that same linguistic Rubicon even earlier? His brain casts are all we have to go by, but their subtle bumps and dimples are certainly intriguing. Perhaps he did have some form of spoken language: a few hundred nasal sounds, say, for those aspects of his hunter-gatherer lifestyle most crucial to him. In any event, by 1.6 million years ago, with *erectus*, the evidence is persuasive.

It is no accident that the first handmade tools and the first glimmerings of a spoken tongue should evolve at roughly the same time. In order to consciously manipulate your environment, you must first have a brain that can manipulate mental models of objects that are perceived. Once you can see the world piecemeal, and consciously isolate and focus on its components, the next step (at least, so it seems with hindsight) is straightforward. It is to pair off each object (or action or emotion) with a symbol—a symbol that can be vocalized in some reasonably predictable, agreed-upon way. From that stems immense survival value, especially for hunter-gatherers who rely upon close collaboration with their fellows.

If you can communicate your thoughts and ideas more

effectively, by some form of speech, then you have established a bridge (albeit a narrow one) between minds. The bridge works because all humans, apparently, share a common view of the world, or at least a common subset of experience. Providing you adopt some standard audible protocol—the words of a language that others also understand—you can reveal detailed aspects of your personal universe to your fellows. Then it is as if you and your family and comrades were part of a single supermind. Each brain may still be physically attached to an individual, but its resources and experience become available to the community.

The coupling between different brains, achieved through a spoken language, is analogous to the connection between neurons inside a single brain. Obviously, the neuron-to-neuron electrochemical coupling works on an altogether different scale and level of intimacy than the brain-to-brain audio link. Yet the two forms of neural connection—intracephalic (within an individual brain) and intercephalic (between different brains)—are very much part of the same evolutionary process. That is, they are both involved in the rise and expansion of consciousness.

Members of other species, of course, also "talk" to one another in languages of varying sophistication—languages that may be based on sound, display, gesture, or, at the lowest level, unconsciously produced chemical messages. In fact, long before man or mammal appeared on the scene, there were primitive creatures circulating primitive thoughts both within their own neural circuitry and between individuals. It is simply that with man the quality of these mental exchanges has increased enormously.

A close parallel exists between all this and the architecture of computers. Like a brain, a computer contains a densely packed mass of interconnected circuitry. Like a brain, a computer can move information around internally at great speed. Computers may also be joined together to form a multiple-processor network. And just as different brains can link up through the use of natural language, so computers can communicate along data highways (generally of much higher capacity than a human speech channel) by means of a standard communications protocol.

To turn the analogy around, a community of linguistic brains is like a distributed network of computers. Working together the brains can achieve far more—handle larger problems, solve more problems at once, and achieve greater insights—than they could by working in isolation. But a network of intercommunicating brains amounts to far more than just a powerful data-crunching machine. Unlike a contemporary computer, a human brain is conscious, and even self-conscious. So a networked community of brains introduces the possibility of a new level of consciousness—a *collective* consciousness in which the whole is much greater than the sum of the parts.

To bring about this new growth of awareness was not easy, however. Depending as it did on the acquisition of speech, it was bound to be gradual—painfully gradual. There were no grammar classes for ape-men. It was simply that improved vocal communication was likely to raise a group's prospects for survival. So a band of hominids that by genetic mutation acquired a brain and voice box slightly more suited to language might be able to communicate better internally, and so enhance its food-gathering efficiency, its defensive strategies, and so on. At the same time, as the population grew and intergroup competition became more intense, the survival chances of less communicative rivals would decrease.

And there were surely other factors at work helping to promote language skills. For instance, youngsters in a linguistically talented group would have more to learn in order to function successfully as adults. Those born with brains (and throats) capable of assimilating all the group's language would tend, in time, to become dominant and hence, presumably, more attractive to the opposite sex. So, these "smooth talkers" would incline to be the ones who propagated their genes, while their more tongue-tied contemporaries fell by the evolutionary wayside.

The point is, there were multiple pressures, both from within and from between hominid groups, toward the evolution of more sophisticated spoken language. And that set up a feedback loop of impressive complexity and extent.

To speak better, you need a better brain. And it is not just a question of tacking on a Wernicke's area here and a Broca's area there. These new outgrowths of the brain have to be fully absorbed and integrated with the rest of the neural machinery, so that simultaneously there are many developments throughout the brain.

To speak better, too, you need a skull base that is more flexed and arched. And that may have been a decisive factor in the sculpting of a more delicate, humanlike skull and body posture. For if you alter one part of the anatomy, you must at the same time alter others because the whole structure has to be brought back into balance. So, for example, if you trim down the jaw and make it of thinner bone, then you must reduce the size and weight of the brow ridges, the cheekbones—indeed the whole skull and skull-support system; otherwise stresses are set up that put the individual at a survival disadvantage. We see that this acquisition of language may have had remarkably far-reaching effects. The very shape and size of the teeth may have altered, in part, to make speech more articulate.

And these are just the physical consequences of the development of speech. The social impact would be no less dramatic. Faced with the problem of how to educate youngsters to a higher standard, evolution turned to the ploy of neoteny. The characteristics of the infant—inquisitiveness, an openness to learn, a half-playful attitude to life—were retained into adolescence, and eventually into adulthood. And as the adult took on more childlike qualities, so the effect riffled back and the newborn became more and more embryonic in character and appearance; its brain, more receptive, was less driven by instinct than ever before; its demeanor one of total helplessness.

Look at a group of chimpanzees or gorillas today, and you can see the effect these tendencies have, even though they are so much less pronounced in the ape. Youngsters who are helpless—and helpless for a number of years—need close attention. They may have left their mother's womb, but now they must spend a much longer period in the collective womb of their extended family. That exerts a powerful binding and stabilizing force on the group. It nurtures a feeling of identity, of protectiveness. And, perhaps most significantly, it encourages the growth of a social hierarchy, with leaders, subleaders, young pretenders, minions, and the rest.

Many powerful agencies then were in operation, stretching and shaping mankind. There were new environments and new opportunities born of new tools and strategies. There was the dawn of language and all that that made possible. There were inter- (and, doubtless, intra-) group conflicts; a rising, shifting, colonizing, diversifying population; new social orders. And all of these factors were interacting, feeding upon each other, in a melting pot of dizzying complexity.

Little surprise, then, that there was rapid change, that man sped from primitive *habilis* to near-modern *sapiens* in just a million and a half years.

By 300,000 B.C., archaic *Homo sapiens* was abroad across all of the Old World. For the next 200,000 years, he altered modestly in the shape of his face, becoming a little less rugged, perhaps to accommodate his increasingly refined speech. He was a prolific stone toolmaker. But still he was no more than that—a strict utilitarian.

By 100,000 B.C., man *was* man. In every essential detail, he was physically what we are today. But still he persisted with his tried and trusted stone tools, creating, apparently, nothing that was not essential.

Tenacious and adaptable he was in the extreme. If he had not already proved that by surviving the dry savannah of his ancient homeland, and pushing out of Africa into remote parts of Europe and Asia, then surely he demonstrated it during the ice ages. In four great waves, the northern glaciers came and went, leaving their mark on all the globe. But while other animals fled before them, or perished, or specialized to an overly high degree, man held firm and adapted once again.

It was to be the final humanizing process. Forged in the arid heat of the savannah, man's mettle was now tempered and tested by the ice-age cold. The freezing conditions made him rely more than ever upon his wits, upon his ability to exploit multiple aspects of his environment. So, where he could, he took to caves as a shelter from predators and the biting wind. He became a clever hunter and butcher of the big tundra-dwelling mammals—the mammoths and their ilk. He lit fires, made clothes of furs and skins. And he survived, even prospered, in this hostile world.

But there is much we do not know about the final metamorphosis to modern man. The details of how our own subspecies, *Homo sapiens sapiens*, emerged are far from clear. Did we evolve from an older stock of *erectus* who invaded Eurasia up to a million years ago and who slowly changed into the archaic form of *Homo sapiens* by about 300,000 B.C.? Or are we the spawn of a later invasion of anatomically modern men who sprang up in the heart of Africa and marched on Europe and Asia perhaps as little as 100,000 years ago?

The truth may be a hybrid. If modern men first appeared in Africa, say around 200,000 years ago, they may well have crossed into Eurasia by way of the Levant (the region bordering the eastern Mediterranean in present-day Israel, Lebanon, and Syria). In fact, skeletal remains provide good testimony of early modern humans in the Qafzeh and Skhul caves of Israel, dating back 90,000 to 100,000 years. By at least 43,000 years ago, populations of modern people had reached southern and central Europe. Yet there is no doubt either that, while this was going on, the earlier wave of African hominids—the erect men—had also been evolving in their new, northerly home. We have the skulls and other bones of *sapiens archaic* dating back to 300,000 B.C. to prove it. And we have, too, the remains of another curious fellow—Neanderthal man.

The case of the Neanderthals surely ranks among the more intriguing in anthropology, and it provides perhaps an opportunity to see what happens when two almost equally intelligent species meet within the same niche. It may have been the last in a long line of such encounters that shaped human consciousness. At the very least, it is a good mystery story—and it may even be a murder mystery.

Sometime around 120,000 years ago in Europe, the first bones of Neanderthal man appear in the fossil record. And they are remarkable bones. They tell of a human being (his usual classification is *Homo sapiens Neanderthal*) who evolved along a slightly different line from ourselves, either directly from *erectus* or as a spin-off of *sapiens archaic*. His remains show that he must have been large and muscular, capable of speed and power beyond the reach of today's best Olympic athletes. We know, too, that at this time and in the regions where the earliest Neanderthals have been found, Europe was enjoying mild temperatures and was blanketed by forest. The teeth and other remains suggest that Neanderthals competed among themselves for access to dense stands of fruits, nuts, and other plants. Compelled to defend private foraging grounds, this otherwise imposing hominid was prevented from pursuing (as other men surely did at this time) seasonal, highly mobile game or from living in large groups. As a result, his social growth was stunted. And we can see in his tools, also, that he had fallen behind his contemporaries. Neanderthal artifacts—the so-called Mousterian technology—are almost exclusively made of crudely flaked local quartz and quartzite.

A picture begins to emerge of a brawny, perhaps slightly dull-witted creature. But then we look closely at a

Neanderthal skull, and suddenly we are brought up short. Because it seems that this puzzling creature had a brain that was actually larger than our own.

Now, that size difference may not be so significant for, in fact, there is a very wide variation in the brain sizes of normally functioning adults today. To make one extreme comparison: the brain of the novelist Anatole France weighed only 39 ounces against the 78 ounces of Lord Byron's brain. Such differences in brain size can apparently be accommodated without any obvious effect on intellect (providing the brains are structurally similar). But what we do not know is why Neanderthals seem to have had consistently larger cerebrums than ourselves. Nor can we be sure how advanced Neanderthal man was in ways that would leave no material traces.

However, by far the most mysterious question surrounding these big-brained cousins of ours is their disappearance. At some point between 40,000 and 30,000 years ago, the Neanderthals became extinct. And that date is suspicious, for it follows closely on the arrival in Europe of modern man.

Around 70,000 years ago, the European climate cooled at the onset of yet another glacial advance from the north. Apparently, it proved a little too chilly for some of the Neanderthals, because their fossils from shortly thereafter have been unearthed in the Middle East. They had sought out warmer lands, where the plants they depended upon still grew in abundance. But in that quest they had found something else that was less welcome—groups of anatomically modern men.

We can only conjecture what happened next. Perhaps the Neanderthals made for low-lying regions containing the familiar groves of fruit- and nut-bearing trees. If so, they may have driven out the physically weaker modern humans and forced them to concentrate in larger groups more suited to hunting mobile herds of antelope and other game on the nearby stretches of savannah. But then, between 64,000 and 32,000 years ago, the world climate cooled further. The savannahs of the Middle East broadened, while all across Europe spread pine forest and tundra. These are exactly the conditions that favor migrating game animals and widely dispersed seasonal plants. It meant that initially in the Middle East the modern hunting humans now had a tremendous advantage, while the Neanderthals may have been wiped out. If this is really what happened, then the effect may soon have spread north and west, so that in time the Neanderthals everywhere were on the wane. It is not hard to believe that at least some of them, beleaguered, weakened from lack of food, faced with a technically superior and better-organized rival, were overrun and slaughtered by the ascendant moderns.

So now the winnowing process was complete. The vying species and subspecies of man had been reduced to just one survivor: the animal that is us. (Perhaps, ultimately, it had to be so. Perhaps there was only room on this planet for one intelligent species. That may even be a universal truth. For as intelligence dawns, it may ruthlessly, if not maliciously, slay all competition until it stands alone, unchallenged. If this is the case, then any future encounters we have with advanced extraterrestrials should prove especially interesting—and maybe a little risky. If it is the tendency of nature to admit only one form of high intelligence to any given niche, we cannot assume that the "galactic niche" will eventually be occupied by ourselves. It may indeed already be taken.)

But there is an alternative way to view human—and sentient—evolution. It is to see all the varied hominid groups as having played their part in the evolution of the human race, exploring different avenues and possibilities for survival. What emerged at the end was simply an amalgam of everything that worked best. And certainly, through crossbreeding over the ages, we are a composite creature with genetic dashes of all the long-dead hominid groups.

However, we have not quite brought the story of modern man's evolution up to date. A vast gulf separates the present day from the point at which Neanderthal disappeared. And this is reflected not in our physical evolution but in everything else that humans have achieved in that time.

Man's physical descent from the apes took roughly 5 million years. And yet the cultural and technological transition from the age of stone tools to that of the United Nations, electron microscopes, and genetic engineering took less than 40,000 years. That is an acceleration by a factor of over 100. And what is more, the switch to this new, accelerated mode of evolution was remarkably abrupt.

It happened 35,000 years ago (or possibly a little earlier) in Europe, and marks the beginning of the Aurignacian period. It signals the boundary between the end of the Middle Paleolithic and the start of the Upper Paleolithic. And it involves the people we call the Cro-Magnons.

What occurred was nothing short of a dazzling social, cultural, and technological revolution—an explosion of new ideas and artifacts. For the first time of which we have any clear record, men consciously created designs, patterns, stylized representations of animals, and ornaments, which would grace any contemporary art gallery, using a fantastic variety of materials and media.

The stone tools are still there, of course, at the practical core of the culture, and they have become magnificent in their precision. Indeed, displayed in museums today, these finely shaped pieces of flint, quartz, and countless other rock types are superb works of art quite aside from being instruments of pragmatic intent.

The Cro-Magnons also fashioned tools of antler, bone, ivory, and even marine shells. They were wonderfully innovative, not only in the materials and methods they used, but also in the range of forms they made and the speed at which they adopted new techniques.

The tools are impressive; they would delight any craftsman skillful enough to make them today. But it is the other products of the Aurignacian—the objets d'art—that are more profound. It is these that reveal the true mind-state of their creator, for they are not constrained by the need to be practical. They are abstractions: frozen images and impressions of the Cro-Magnon mind. Yet, despite their seemingly capricious nature, they were surely made for some important purpose.

Pendants have been found, for instance, made from the pierced teeth of animals—usually predators, such as the wolf, the hyena, and the bear. It may be that in choosing these as ornaments the Cro-Magnons were trying to evoke some of the powers of those species. That is a very advanced form of symbolism, whereby a part is made to stand for the whole, and the powers of the part are taken to flow into the wearer. The need for ornaments of power to mark status within the increasingly complex Cro-Magnon society may indeed have been the instigating factor behind this whole new mode of thinking.

Other decorative artifacts also seem to have been conceived as metaphors for real objects. At Volgelherd, for instance, in southern Germany, a 32,000-year-old horse was found, carved of mammoth ivory. It is a tiny thing, two inches long. And what is so revealing is that it has obviously been much-used, carried perhaps in a pouch, or handled over and over again, as if from that close contact the owner hoped to extract some special, privileged magic.

From symbolic art (if that is how it began), the Cro-Magnons quickly went on to apply their new, highly visual way of thinking to more practical matters. It suddenly became possible for them to transfer all manner of properties, such as "pointedness" or "barbedness," from natural contexts to technological ones. Having abstracted the concept of, say, "barbedness" from nature, the Aurignacian toolmaker might then have been able to visualize alternative designs for a barbed spearpoint. That type of thinking is at the very heart of invention. During the Industrial Revolution in Europe, for instance, all the great inventors thought in images. Indeed, our continuing rapid technological development today rests upon the forming, manipulating, and sharing of images that began in the last ice age.

And so we come almost to the dawn of historical times. Very clearly now we see man in two quite different ways: as an integral part and product of nature and, at the same time, as a creature able mentally to draw a sharp distinction between itself and the objects of which it perceives nature to be made. Man had begun to watch and study his surroundings with keen interest, to manipulate them to his own end. He became like a director (and, in time, even a scriptwriter) of nature's play, rather than an actor locked helplessly into a particular role. That was the basis of his success. Yet it was the beginning, too, of a most fundamental divide—man and nature, inner world and outer, mind and matter. It is the dichotomy that lies at the core of the human condition.

4

A Parting of Ways

We live simultaneously in two worlds, because each of us has two brains. And although those two brains—the right and the left cerebral hemispheres—are connected, and work admirably in concert, they nevertheless function in quite different ways.

The left hemisphere excels at analytical, sequential thinking. It is the rationalist within us, the reductionist. Largely through the workings of the left brain, we break the world down into components, see it as a mosaic of bounded objects: objects which include ourselves, our bodies, our brains, and our very thoughts at this instant.

By contrast, the right brain deals principally with holism. It is the massively parallel computer to the serial processor of the left: a specialist in perceiving spatial relationships and the nature of the total configuration, rather than individual features or elements. If the right brain has any sense of identity at all it is as a mere conscious oneness with the whole of existence. The right brain puts up no barrier between itself and the undivided universe. Yet, only an inch or so away, the left brain calls itself "I," sees itself as sharply distinct from what lies outside, and broods incessantly over its eventual demise.

Two brains, joined like Siamese twins, snug within their bony cave. But the relationship between the pair is uneasy. The left can talk (internally and externally)—and does so, with little respite, throughout the day. It can talk because it has both the key language centers, Wernicke's area and Broca's area. And that is in keeping with the rest of its talents: objectifying, identifying, classifying, and—through language—labeling. The left is not happy until it has put the world into a million different boxes and tagged every one. But even then it has the feeling that the task is undone, that something has escaped its attempts to reduce the universe to its most fundamental parts. That is why, in the end, science and all other rational pursuits are never completely satisfying. In taking apart and analyzing, they lose the essence of the whole. The truth is that the universe is an unbroken totality. And though immeasurable benefits may come from looking at the world in pieces, they will never include an appreciation—still less, a direct perception—of the unity of nature.

Why should it be that we experience the world in these two quite different ways? In fact, the ability to classify objects as well as to globally perceive an unbroken whole is something that other animals have, too. Chimps, for example, categorize colors in the same way that humans do. You can teach a chimpanzee words, in sign language, that correspond to the wavelengths of the primary colors. And what you find is that the chimp consistently uses the same color terms as every human culture on the planet.

More dramatically, it can be shown that pigeons (that much-maligned bird of the psychologists) classify objects in the world much as do you and I. A series of experiments has been carried out in which pigeons were presented with two tiny screens, onto each of which was projected a photographic slide. On one screen an image appeared containing the target concept to be learned, while on the other a picture was shown that was devoid of such a reference. So, for instance, if the concept was water, appropriate images would include oceans, lakes, puddles, and raindrops, at every conceivable magnification from closeup to long-range. The pigeons were rewarded with tidbits if they pecked at the correct screen. Remarkably, once they had learned a certain concept, such as water, yellowness, mountain, or roundness, they would respond unerringly when tested with hundreds of pictures they had never seen before. They could even select pictures demanding a power of discrimination far beyond that of the experimenters. For example, the birds could spot a particular individual among thousands of faces in a crowded football stadium.

That comes as something of a shock, and even an affront to our intellect. "Lower" animals capable of identifying and classifying objects in the environment in the same way as we do? Well, in fact, it should not be so surprising. Classification is fundamental to survival in the real world. Unless a creature could categorize phenomena into more general types, it would have to treat every one as unique. Then the creature would have no idea what was good to eat or what represented a threat. Before it had time to establish a thing's credentials, it might be too late—the food would have gone or the predator would have struck.

One crucial fact emerges from this: the world cannot be random. If it were, it would be unpredictable. The categories that we (and other animals) recognize must be based on natural, recurring physical properties of the universe. That may seem obvious, but it is only because we are used to looking for—and finding—order, consistency, and predictability in our surroundings. Indeed, it is far from obvious, a priori, why the universe should be anything other than totally random and intractable. Randomness always seems so much easier to achieve in everyday life than organization; leave a room to its own devices and the point seems well illustrated. How could nature, without any conscious effort, do a better job of keeping its own house in order? And yet, against all the odds, the world is patterned. And what is more, over time, these patterns have become increasingly elaborate and exquisitely organized to the extent that now the universe has created, from within its fabric, creatures of such extreme complexity that they can discern this order and classify it. Remarkably, elements of the universal classification have become capable of classifying the universe—including themselves.

But this raises an important question: Do we always see objects and relationships that are *really* there, independent of our mental selves? How can some thing, or some class of things, or some connection between things, be said to exist without a sentient observer to perceive them? Are we not, in fact, as much inventors of the order we see as discoverers? In some curious way, the two seem to go hand in hand. Our brains make sense of the data they receive: they impose or invent order. And yet the basis for that perceived order must somehow already be there. It is one of the fundamental mysteries of nature, this dichotomy between what is given and what we, with our minds, create. We owe our very existence as a species to our ability to delineate patterns. We can even see patterns where none exist—the faces in a sun-lit curtain, the Greek heroes and monsters among the stars. What else might the human mind be recognizing that is not really there? And what, in any case, do we mean by "real"?

Everything we see, everything we know, enters our consciousness through the mind-brain mechanism. Everything—the "outside" world, your body, and, curiously enough, the very substance of your brain—is part of the experience that the brain creates. In the words of the physicist Sir Arthur Eddington: "Mind is the first and most direct thing in our experience. All else is remote inference—inference either intuitive or deliberate."

That is a sobering (and slightly disturbing) thought. We slip so easily into the habit of assuming that what we see and feel in our minds is what is actually going on outside ourselves, beyond the portal of the senses. After all, we are only inches away from the borders of this seemingly familiar land. But there are no colors *out there,* no hot or cold, no pleasure or pain. Although we experience the world as a series of sensory objects, what actually comes to our senses is energy in the form of vibrations of different frequencies: very low frequencies for hearing and touch, higher frequencies for warmth, and higher still for vision. These radiations carry no subjective value whatsoever. They are simply part of the energy scheme. And even in that much-diminished role they are denied anything but the most phantasmal existence by modern physics.

The radiations we pick up trigger neural codes that are made by the brain into a model of the external world. Then this model is given subjective value and, by a trick of the brain, projected outward to form the subjective world. That inner experience is what we habitually equate with external objectivity; we get what we see—or so we believe. But it is *not* objective. Even when we look at that most pragmatic of scientific tools, a measuring device—a clock, a Geiger counter—we are watching only a subjective mental model fabricated within our brains.

The problem is not laid to rest, however, by this realization. We have also to consider that our brains and everything that happens inside them are part of the reality of the universe. By any reasonable definition, nothing exists that is not real. Does that mean, then, that unicorns are real because you can think about them? No. But it does mean that your mental model of a unicorn is real—as real as any hornless white horse that ever lived on Earth. The mental model may be an entity of an entirely different caliber, or be on a different physical level, from a horse you can actually ride, but that makes it no less "real."

All of perceived reality is a fiction. And yet, reassuringly, you and I and the one hundred billion other humans who have ever lived seem to perceive a similar external world. Or to put it another way, we have no grounds for suspecting otherwise. So there is more than a mythical basis to our private fictions. We are part of and participate in a coherent, highly patterned reality.

But we have two brains. Or, at least, we have two quite distinct brain hemispheres: one a specialist in holism, the other in reductionism—a mystic and a scientist. And that is highly significant, because not only is the brain a part of the universe, but its structure and its operation lend clues to the nature of the universe. So when we see that the brain is divided into two halves, each with its own characteristic mode of thought, then we are looking very deeply into the way reality is constructed.

There are, it seems, two essential, complementary aspects to this cosmos. One is the whole: the indivisible, continuous, fluid unity. And the other is the interrelationship of parts, in space and time. The brain has developed in such a manner as to take account of this dual aspect. And the reason is that the perception of both the whole and of its components is essential to survival, yet cannot be achieved simultaneously. You may be a mystic or a scientist, but you cannot be both at the same instant of time.

It would be interesting to know when this curious split between the left brain and the right brain began. However, that is very hard to say based only on fossil evidence because, from the outside, there is virtually no difference between the two hemispheres. The morphological symmetry has been largely preserved even in modern human brains, which are the most highly "lateralized" on Earth. Our ancestors' skulls offer even less scope for spotting irregularities, and any results are bound to prove controversial.

Fortunately, there is no need to strain our eyes peering into petrified craniums. A great deal of research has been done on brain lateralization in living animals of very diverse species. And it now seems that asymmetry connected with a large number of different functions probably exists in every type of mammal and bird—and, indeed, in most if not all vertebrates. That suggests the origins of the divide must be quite ancient. The question is, why? Why was there the need for this separation—and so long ago?

A need to be globally aware of your surroundings is vital for any reasonably complex animal. It does you no good to be able to focus like a laser beam on your next mouthful of food, if a predator, outside the beam's glare, is about to make a mouthful of *you*. The most basic function of any brain is to proffer some idea of what is happening all around you, without particular attention to any one thing. That is a holistic awareness—taking in the big picture. But, in turn, it is not much use seeing everything if you cannot zero in on the particular aspects of everything that affect your well-being. When dinner or danger shows up on your holistic radar screen, *then* you need to be able to switch immediately to a more focused consciousness—to turn on the laser.

All animals, at least as far back as fish and amphibians, would seem to need these two types of mental outlook. The point is, however, they cannot be handled precisely by the same brain circuitry. In holistic, or unfocused mode, the brain is accepting multiple inputs and treating the scene it builds from these as a single entity. In computer terms, that is a parallel or array-processing task. But the other mode—the focused state—isolates the target of interest and tracks it single-mindedly and sequentially. That is a job for a serial computer, working step by step. So there are two very different processing tasks requiring, to some extent, different neural pathways and neurophysical structures—a raison d'être for the divided brain.

We should be careful, though, not to overstress the poles-apart nature of the right and left hemispheres. In fact, each hemisphere does itself consist of many different regions and subregions that perform in markedly varied ways. There are also millions of connecting nerve fibers comprising a central body called the corpus callosum, which is a sort of grand data highway between the two brain halves. As a result, anything that one hemisphere thinks about, the other hemisphere learns with very little delay. It is also true that both hemispheres do a lot of the same routine jobs. For instance, each hemisphere is in charge of sensory input and motor control associated with one half of the body. And it would never do, say, for one leg to walk rationally, step by methodical step, while the other was somehow poised in a state of mystical rapture.

We can only push the hemispherical differences so far. But, in fact, even with respect to the functioning of the two sides of the body there is something of a holistic-rationalistic divide.

Curiously, each half of the brain is responsible for the visual, auditory, and tactile space associated with the opposite side of the body. So, for instance, the right visual cortex (also, strangely, at the back of the brain) is wired up to deal with the left-hand field of vision. That is to say, it is connected, not exclusively to the left eye, but to the left-hand portion of the retinas of both eyes. The same type of cross-wiring applies to the ears and the sense of touch (not, however, to smell, the brain pathways of which may be more ancient).

Some dramatic consequences follow from this eccentric neural circuitry—consequences that become clear when a person suffers an injury to just one cerebral hemisphere. Damage to the right side of the brain, for example, leads to the clinical syndrome of left "hemineglect." Patients with this condition disregard events occurring in the left half of the space around them, and in severe cases it is as though that side of the world no longer exists for them. They may shave, groom, or dress themselves on the right, leave food uneaten on the left side of the plate, write only on the right half of a sheet of paper, and, when asked to mark the halfway point on a line, place the division well to the right of the true midpoint.

That is remarkable enough. But if we consider a situation in which this spatial division is combined with the functional separation of the two brain halves, the results can be quite astonishing.

Experiments carried out with chicks shed some preliminary light. In birds, in turns out, the optic-nerve fibers cross over completely, so that each eye sends its messages solely to the opposite side of the brain. By covering up one or the other eye, it is possible to see if this has any effect on behavior. And the strange result is that it does—but only in the case of male chicks. In visually discriminating food from pebbles, male chicks whose right eyes are covered can learn the task perfectly well. However, chicks with patches over their left eyes perform poorly. It seems that the left eye (and right hemisphere) takes into account the position of an object in space, but is not so concerned about its detailed characteristics, such as color and texture. Instead, the left hemisphere attends to these details. Nor is visual discrimination the only function lateralized in the chick's brain. A male chick, for example, responds to new objects much more strongly when it sees them with the left eye. And, most bizarrely, a male chick treated with testosterone will copulate when seeing out of just its left eye, but not out of just its right. All of which leads to two tentative conclusions. First, that behavior depends upon which sensory field (right or left) is being stimulated. And second, that the asymmetry of the male brain (possibly due to the prenatal effect of testosterone) is greater than that of the female's. Do these findings extend to humans?

We take immense pride in our superior intellect, even though, as individuals and as a race, we played no conscious part in its evolution. The truth is, however, that many of our vaunted mental powers differ only in degree from those of other animals. Mental evolution, like physical evolution, has been a gradual, incremental process. And, as embryos, we clearly reveal our ancestry by recapitulating that process at lightning speed before taking it an impressive stage further with the ballooning of the human cortex. We may, comparatively, be mental giants, but it is only because we stand on the shoulders of what came before us.

Other animals, too, can recognize objects, as well as have a general, holistic awareness of what is going on around them. They can also recognize classes of objects, and even, in the case of the higher mammals and birds, manipulate mental representations of those objects. There is no reason to doubt that they enjoy a rich inner world, and experience both focused and unfocused consciousness. We share with them, to a greater or lesser extent, the same fundamental cognitive machinery and processes of inference. The plain difference is, we have taken things further. We can aspire to higher levels of mental abstraction that involve being aware, not just of objects and their properties, but of "ourselves" and of ways in which we can influence the future of our surroundings.

As our brains have evolved, so they have become more and more strongly lateralized. And, in particular, two of our special human talents seem to have acted as a powerful stimulus to this left-right polarization. The first is language, which requires a stepwise, logical formulation and so found a natural home in the left hemisphere. The second is toolmaking.

Neither of these talents, of course, sprang up from nowhere in our hominid predecessors. Their roots stretch back tens—perhaps hundreds—of millions of years to the first creatures that had a dim sense of category and causality. It was simply that in humans the time and conditions were right, the eons of preparations were complete, and so these latent, high-level survival skills could come to fruition.

As for language, the great apes are not far from possessing it even in their natural state. It may be that they already have in their brains the shadowy beginnings of speech centers. But in captivity, taught by creatures who already have an advanced language, there is no doubt about the apes' linguistic potential. There are gorillas and chimpanzees in research centers today who have vocabularies of several hundred words and can sign their feelings and intentions as clearly as a three-year-old human child. Nor are they far, in evolutionary terms, from being able to make their own tools. Indeed, their closest living relative—man—took less than three million years after branching from the ape line to become a stonemason.

The making of tools, like the formulation of language, involves ordered and rule-governed activities. These are activities for which the left hemisphere was already partly specialized. So it is not surprising that the rationalist left side should be the one to take up the mantle of technologist. That would also explain why man is one of the few living things (rather incongruously, parrots are another) to show a preference for one hand over the other. About 90 percent of humans are right-handed; that is, they favor for precision work that side of their body controlled by the left brain.

Steadily, from about two or three million years ago, man's organ of thought became increasingly bifurcated. This is particularly true of the human male, because the polarization of the right and left hemispheres seems to be more pronounced in men than in women. A hint of such a gender difference has already been seen in the chick experiment, and may apply to many higher animals. Research on humans seems to show that in our species, too, the sexes differ in the relative importance of left- and right-hemispherical processing.

Again, consider language. Up to now, we have oversimplified how this is handled by the brain. In fact, both left and right hemispheres are involved, though the left predominates. In most individuals, the left side sequentially analyzes, or parses, the sounds and grammar of the language. It also produces speech and stores many words (its so-called lexicon). The right side is more concerned with recognizing the shape and form of the words, and has a smaller lexicon. Although both men and women make the same asymmetrical use of their hemispheres to comprehend language, some evidence suggests that, when speaking, women use their hemispheres more symmetrically.

In tests, women score higher overall than men on problems calling for both verbal and visuo-spatial reasoning (right-brain functions), while men on average fare better at solving numerical questions. There is also the general observation that women tend to be more perceptive and responsive to the emotions of others. And, again, it is the right brain that specializes in the decoding of emotional significance.

Concerning this last point it is curious that, when cradling a baby, 80 percent of women hold it against the left side of their bodies irrespective of whether they are left- or right-handed. (Female chimpanzees and gorillas show the same bias.) There is a possibility that this may be to position the infant nearer the mother's heart, which is slightly displaced to the left, so that the baby will be soothed by the sounds of the heartbeat. But there are two main objections to this theory. First, the "beats" are actually the sound of the closure of the heart valves, and these are located more or less centrally, under the breastbone. And second, the preference for cradling a baby on the left has been observed even in African mothers who carry their infants on their backs all day while working.

An alternative explanation invokes the divided brain. That is, by left-cradling, a mother is able to monitor the baby with her left visual and auditory field—the field linked to the side of the brain best able to interpret emotion. On top of this, because there is evidence that lateralization also applies to facial expression, it may be that this way of cradling offers an advantage for the infant, too. Namely, the baby can see the left-hand, more emotionally expressive side of its mother's face.

As a species—but especially among the males of our species—we have shown a tendency to turn away from nature's expressive side. The evolutionary momentum that gathered pace from the time of habiline man, that brought us tools of growing sophistication and gave us speech, has involved especially the left sides of our brains. We have become supreme rationalists, schemers, and engineers. We have probed and probed the inner grain of the world that has nurtured us and have exploited it relentlessly for our own ends.

Following the techno-cultural breakthroughs of the Aurignacian, and the key discovery that natural properties can be transferred to other contexts, our progress became exponential. And, of course, it has not all been at the expense of the right-brain experience. Because when finally our technology gave us stable city-states, and we could farm our own crops and animals, and we became relatively safe and secure, then we had the luxury to enter the life of the mind. Artists, musicians, poets, philosophers, and priests could indulge at leisure in their significantly right-brain activities, so that in some sense the right brain profited from the industry of the left.

Many religious world models display an intuitive knowledge of left- and right-brain functioning. In Taoism, for instance, there is a male principle, known as *yang,* that is clear, rational, and dominant. At the other extreme is *yin,* the female force, characterized as complex, intuitive, and yielding. Life, according to Taoism, should be a blended harmony of the yin and the yang. It is this harmony we seem to have strayed away from. In Western culture the left side of the brain is the more active and the male principle dominates, which may explain why the West is so technically advanced and yet in some ways is so spiritually impoverished.

In the East, religion is natural philosophy. It is an unhurried, holistic, timeless experience, centered in the right brain. We can even demonstrate that it is such. With electroencephalographs, we can monitor the neural activity of Zen masters and Buddhist monks in meditation. And, on the brain-wave plots, we can see a great active area on the right side of the skull. The mystic experiences it directly; we merely watch the brain trace. And there is the point of separation, the parting of the ways.

Our brains have evolved so as to see the world in two different, complementary but also mutually exclusive ways. Each of us, figuratively speaking, has the East and the West, the female and the male principle, in his or her head. But usually one or the other has the ascendancy. Either we are too concerned with rationality and so, from the Taoist viewpoint, fall out of harmony with nature, or we are too introspective and fail to achieve materialistic growth. Both mental modes are apparently essential to human consciousness and so ought to be brought more into balance. Indeed, there is evidence today, both in our physical sciences and our growing concern for the well-being of the planet, that this is already beginning to happen.

It has taken billions of years to fashion human brains. Now, through the workings of these brains, nature is able to pose the most profound questions about itself: How was the universe made? Why does it take the particular form it does? And, most intriguingly from our personal point of view, is there a role for consciousness beyond simply enhancing the survival chances of its owner? In other words, does man's existence and awareness have any wider meaning?

There have been many attempts over the ages, by theologians, mystics, philosophers, and scientists, to furnish adequate answers. One is a basic recurring theme in Hindu mythology. It is that the world—the cosmos—was created by the self-sacrifice of God. God became the world, which, in the end, becomes God. The concept is called *lila,* the play of God—the world seen as the stage of a divine drama. Because lila is dynamic and rhythmic, *maya* is in continuous flux. Maya is the illusion of believing that the shapes, structures, objects, and events around us are "real," that they are more than the artifacts of our measuring and classifying minds. And then there is karma, the dynamic force of the play. Karma is the whole universe in action, where all is dynamically interconnected. What we have to do, according to Hinduism (and it is a common goal of Eastern religions), is to realize within ourselves this underlying unity and harmony of nature. We need to wake up from the dream of maya and see that we are not separated from our environment; indeed that, ultimately, "we" are an illusion. It means experiencing, intimately and directly, that everything, including ourselves, is what the Hindu calls Brahman—the whole.

Until the first quarter of this century, this type of human-centered, "participatory" cosmology would have been the antithesis to Western scientific belief. Previously, the conscious mind was regarded by physical science with disdain. Consciousness was an irrelevance; the observer a mere diarist in a universe supremely impartial to human presence. But now we have been compelled by modern physics to regard things in a very different light. As we shall see, we have been forced to concede that not only may consciousness have a purpose, but that it may actually be indispensable to the universe in which we live.

Part II

MATHEMATICS AND MATTER

God has put a secret into the forces
of Nature so as to enable it to fashion
itself out of chaos into a
perfect world system.

—IMMANUEL KANT

5

The Code Within

The universe is predictable—up to a point. And that is just as well, because if it were not there could be nothing as complex as ourselves. Indeed, there could be no evolution of any kind—of life or stars or atoms—because there would be no continuity from one stage of development to the next.

All animals depend upon this predictability of nature for survival, and some make conscious use of it. The Egyptian vulture that persistently throws a pebble at a stolen egg is surely aware that the egg will eventually break. With less sophisticated creatures, however, the situation grows murkier. Do ants, for instance, in some sense "know" how to build nests? Do they have some genuine collective intelligence? Or are they simply part of the familiar pattern of the world, waiting to be exploited by some smarter predator?

As we move up the biological hierarchy, we find animals capable of discerning more and more in nature's predictable scheme. Until we reach man. And then something very extraordinary happens. Through the workings of the human brain, nature begins to understand itself; and more than that, it begins to discover within itself hidden laws—laws, moreover, that obey a remarkable inner code.

Perhaps it stemmed from the splintering of wood or the flaking of stone, for you can only do that so many times before you notice that there is a grain—always in the wood, and often in the stone. Strike along the grain and the fracture is clean and easy; against the grain, the going is much harder. So, in time, unconsciously at first, you learn that nature is not only patterned on the outside, but that it also has a deeper, inner regularity. The same type of stone consistently fractures in the same way. The fracture is man's, but inside a hidden print lies waiting to be found.

That may seem a humble beginning for science. But it lays down the method for all successful future inquiries about the universe: pry a thing apart and lay bare the inner grain. Then, having analyzed the internal structure, we can take a natural form and put it together in new ways.

It took man at least two million years to start to really exploit that idea. Perhaps it was a matter of attitude. Or possibly his visual imagination and technical skills needed that much time to mature. Doubtless the rigors of the last ice age helped to focus his powers. But when at the dawn of the Aurignacian period, 35,000 years ago, man the scientist and the engineer began his research in earnest, he never looked back. Our ancestors started to peer into every cranny of nature that they could, like children who find a secret attic piled high with marvelous treasures.

Quickly, this probing of the world within the world became a habit, a way of life. And it went far beyond—literally, far beyond—the chipping of stones and the almost celebratory exploration of all kinds of novel materials. In learning how to look more closely, man began to notice other types of regularities: the pattern of the seasons, the movement of the stars, the waxing and waning of the Moon. There is some evidence that even before twelve thousand years ago—that is, still during the last ice age—people were making simple calendars with marks on bone and ivory. By ten thousand years ago, though, they had a practical need to. Because ten thousand years ago, the ice age ended, the glaciers retreated, and the brilliant culture that had flowered amid the frozen wastes suddenly awoke to a warm Eurasian spring.

Now the intellectual and practical capacity that had almost run away with itself in the Paleolithic period was unleashed on a more hospitable and yielding environment. Within a hundred generations of the great ice sheets' melting, man had domesticated farm animals, planted crops, and built his first cities. That is a measure of the pace of cultural evolution—and its headlong rush has not slowed to this day.

With agriculture came the need to keep methodical track of the seasons, so there were the beginnings of astronomy. And with settled communities and the growth of trade, matters of business and finance had to be addressed. For both these very pragmatic purposes—the keeping of calendars and the keeping of accounts—some form of reckoning was essential, and so mathematics was born.

Recent discoveries by archaeologists have shed some fascinating light on the stages by which mathematical concepts evolved, specifically in the case of one of the most ancient human societies—the Sumerians. Around 8,000 B.C., the Sumerians were using trading tokens in such a way that it is clear they recognized the concept of number but not its independence from the type of object being counted. For instance, a Sumerian tradesman of this period could count sheep and he could count goats, but he would never think to count the two together. By 3,100 B.C., however, the Sumerians had generalized their concept of number so that it was independent of whatever was being enumerated. They now had separate tokens to distinguish numbers of things from the identity of the things being counted.

The point to emphasize is the very down-to-earthness of mathematics at this formative stage. It was the tool of farmers and tradesmen; it was rooted in the humdrum world of daily human affairs. You match off your thumb with the first of this man's sheep, your index finger with the next, and your middle finger with the next, and so on. And you give a label to each finger—"I" for the thumb, "II" for the index finger, "III" for the middle finger. Next, you realize that not only fingers can be numbered so that if the man has fifty-eight sheep, it is easier to make so many marks in the sand, or to use so many tokens, than it is to employ six people with six sets of hands. You have abstracted the concept of number and the basic operations of addition and subtraction. It does not seem much; yet it is as important a tool in the survival and growth of the culture as the stone chopper was to man's biological struggle.

The interface between mathematics and everyday reality appears sharp and immediate at this point: one sheep, one finger, one token; another sheep, another finger, another token, and you can take away tokens or add them, as you can with your fingers. The tokens—the numbers—are just abstracted fingers; the operations for dealing with the tokens are just the abstracted raising or lowering of the fingers. You make a one-to-one correspondence between the tokens and whatever it is you want to reckon, and then forget about the fingers.

At first, it seems clear from this that mathematics must be somehow already "out there," waiting to be discovered, like the grain of the stone. One sheep add one sheep makes two sheep. Two sheep add two sheep makes four sheep. That is certainly the practical end of the matter as far as the shepherd and the merchant are concerned. But already, even in this most simple mathematical maneuver, something strange has happened. In saying "one sheep add one sheep" we seem to be implying that any two sheep will always be identical. But that is never the case. Physically, the first sheep is never exactly equal to the second: it may be a different size, have different markings. It takes only one molecule to be out of place between the two, and they are not identical. Indeed, because they are in different places they are inevitably not the same on that basis alone. We have extracted a perceived quality to do with the sheep—namely, their "oneness," their apartness—and then merged this quality by means of another abstraction—the process of addition. What does it mean, physically, to "add" things? To put them together? But then what is "putting together" two sheep? Placing side by side, in the same field—what?

All this may seem like nit-picking. But on the contrary, it brings us back to the central mystery—the relationship between the inner and the outer, the world of the rational mind and the world "out there." In the physical world, no two sheep are alike. But, more fundamentally, *there are no "sheep."* There are only some signals reaching the senses, which the left brain combines and then projects as the illusion of a solid, relatively permanent thing we call a sheep.

Like all objects, sheep are fictions: chimeras of the mind. It is our left hemispheres, having through natural selection evolved this skill for extracting survival-related pieces of the pattern, that trick us into seeing sheep, trees, human beings, and all the rest of our neatly compartmentalized world. We seek out stability with our reasoning consciousness, and ignore flux. We shut our eyes to the continuous succession of events if those events seem not to substantially affect the integrity of what we see. So, through this classifying and simplifying approach we make sections through the stream of change, and we call these sections "things." And yet a sheep is not a sheep. It is a temporary aggregation of subatomic particles in constant motion—particles which were once scattered across an interstellar cloud, and each of which remains within the process that is the sheep for only a brief period of time. That is the actual, irrefutable case. But it is not a matter of life and death to know it, or to be constantly aware of it. All such detail in the affairs of the world is irrelevant to human survival; so we have trained ourselves—the left brain has evolved—so as to ignore it, or rather, never to notice it.

And in the roots of mathematics, too, we conveniently overlook this fluxlike essence of our surroundings. Because in abstracting the notion of "oneness," as applied to anything, we are making assumptions: first, that the thing can be treated apart from its surroundings; and second, that there is a quality about it that is permanent and consistent. The "oneness" of a sheep is the same as that of any other sheep—or of a goat, or of a tree.

Often it is remarked that mathematics is precise, aloof, immutable, whereas our physical knowledge of the real world is approximate and always changing. We can see now why this is necessarily so. From the very outset, our rational mind extracts just that which it perceives to be most simple and constant among the messages that enter from outside. Then it abstracts highly purified operations, like addition and subtraction, from the real-world processes of continuous motion and collective interaction. And, finally, it applies these ultimately abstracted operations to its ultimately idealized qualities. So the fact that mathematics builds up like an eternal, unchanging crystal is assured from the start. Indeed, that is its very strength, because wherever you are and whoever you are, one and one is always two, two and two is always four.

We have, then, the beginnings of a new language—the purest of languages—to deal with the most perfectly abstracted qualities of things. There are other languages for handling the more complex aspects of life: the languages through which we communicate our feelings, our interaction with the world, and so on. These "natural" languages embrace the more imprecise, intricate, and emotional aspects of experience, that is, life as it is for most of the time. The words and grammar of our native tongues give voice (both outside and inside our heads) to familiar concepts and actions in the human theater. These speech-symbolized concepts and actions are still abstractions—convenient, survival-oriented products of the left brain. But they try to capture the real world as it comes to us, or as we imagine it to be. As a consequence, natural languages have words not only for the logical creations of the left brain—"water," "carrying," "I," "you," "far"—but also for the emotionally colored, more complex thoughts of the right—"love," "tranquillity," "anguish," "holism." It is a task delegated to the left brain, principally, to do the categorizing and "linguifying" of concepts from both the left *and* the right, even though it cannot directly experience the feelings of its neural partner. But, again, the key point is that all languages represent approximations and simplifications to the one true reality.

Reality is a continuous, all-embracing flux: a universal process that cannot be captured completely in any linguistic form. It eludes description because the very essence of a language is to isolate aspects of the whole scene, and label those aspects. That, of course, is what makes it so useful. We may, philosophically, be near the mark if we insist that the universe is a single, ever-changing, self-perpetuating process, but it does not help practically. It does not help the squad of *erectus* hunters on the plain to corner a lame zebra, parents to teach their children, or you and me to understand a little more of the reality in which we are embedded. Language is an abstraction, a left-brain mock-up of the real world. And yet, for all that contrivance, it proves extraordinarily useful—otherwise, it would never have come about.

Mathematics, too, is a form of language, but a highly specialized one. It starts with the most fundamental abstraction it is possible for the rational mind to make: that an object has "oneness." Nothing could be more purely distilled than that, more stripped of emotional or holistic ties. The same object, as we perceive it, may coincidentally be tasty, lie at the bottom of the ocean, tell good jokes, or, through its exquisite sounds, stir powerful emotions within us. But whatever the object is, by virtue of appearing to us as a separate entity, it has "oneness." That is the clinical point of origin for mathematical thought, and it means that from its very beginnings mathematics has built into it the idea that the world is divided into objects. Furthermore, mathematics accepts implicitly that the quality of oneness of these objects can be considered separately and manipulated in various precise ways—ways which are themselves radical abstractions of the physical reality.

From these elemental building blocks, the whole edifice of mathematics steadily grows. Concept upon concept, it rises from the ground, like the stones that make the walls of a city. Except that mathematics is a technology built, not of the hand, but in the mind. Where are the temples and aqueducts of mathematical man? Where are his telescopes, trains, and skyscrapers? They are nowhere. And yet, in another sense, they are everywhere, for our minds can discern both a concrete, or physical, aspect and an abstract, or mathematical, aspect to all things. The question is, how are these two very different aspects related?

At first sight, it seems that the physical world must

be the more "real," because it is tangible, touchable. It hurts if you bang your head on a physical wall, but no one was ever the worse for running into a mathematical plane.

And yet we have come far enough along this voyage of consciousness not to rush into facile judgments about what is real and what is not. The fact is, there is nothing that is not real if you allow that mental phenomena, by virtue of their undeniable existence, are as "real" as physical ones. In some sense, their reality is even more immediate, more self-evident than the colorless, phantom energy fluctuations arriving at our senses, for it is only the mental phenomena that we actually experience. Everything you have ever known, or felt, or will feel, is mental. Our entire perceived and imagined universe is personal, formulated within our heads, and we can only remotely infer what may lie outside it.

So we need to be careful when considering the relationship between mathematics and the world beyond the senses. Both, in truth, find substance only in our minds. Both, in some way, require the agency of mind to be assembled from the wash of ghostly energy waves reaching our senses.

And yet there is a difference. Whatever mathematics and the physical world may be, they are not one and the same. The qualities that mathematics deals with belong to a higher level of abstraction than the others we perceive in connection with objects. For example, the quality of "oneness" is somehow elemental and consistent, while that of redness is not. Both are qualities that, through our categorizing minds, we filter out from the entirety of what we see and then give a label to. But "oneness" is uncompromisingly absolute, while redness invites judgment.

There is a range of hues that each of us is prepared to put under the umbrella of "red," though there is no consensus about the exact point at which a red becomes, say, an orange or a purple. We talk of crimson and scarlet and vermilion—all of them red—but there is no such leeway in the concept of one.

This difference in "purity" or level of abstraction between mathematical and other properties is, at least, partly due to the way we look at things. Usually we describe colors, for instance, in terms of their appearance—that is, by the physical effect that different types of light have upon us. Look at a rainbow, and you perceive seven distinct colors with only a narrow transition between each. We—our brains and our eyes—have evolved so that we see the spectrum in this clumpy, artificial way. And for a good reason: if your mind can conjure up a sharp difference in appearance, for example, between the fruit that is good to eat and the leaves that are not, then that is a great survival asset.

But there is another way to describe colors, and it stems from the use of technology to probe a little deeper into the nature of things. We learn that, in fact, the brain is playing a (useful) trick; that a rainbow is actually composed of an unbroken spectrum of wavelengths of light, extending from the longest "red" waves to the shortest "blue" ones. Once that is established—once we are aware of this more fundamental property of wavelength behind the subjective quality we call color—then mathematics comes back into play. We can talk about specific wavelengths, and refer to them with absolute, mathematical precision. In other words, we can put numbers to colors, just as we can count a man's sheep. We have extracted from the total, sensory impression the pure, underlying mathematics. And yet in gaining this precision we have lost sight, literally, of the psychical part of the phenomenon. You could spend all day trying to describe redness in terms of the length or frequency of light waves. But if the person listening to you had been blind since birth he or she would be no closer, at the end of it, to sharing the *experience* of redness. Because that experience is precisely (and intentionally) what has been left out of the mathematics.

Mathematics deals with concepts that seem to be invariant between human minds. We may disagree about the relative beauty of a Manhattan skyline and a Scottish loch. We may contest the merits of Brazilian coffee over Kenyan, or of Segovia over Santana. In fact, we may interpret a large part of what we see and feel in remarkably different ways. But that is not true of mathematics. Five is a prime number whether you like it or not. The ratio of the circumference of a circle to its radius is approximately 3.142 whether you are the Pope or the President. Mathematics is a common denominator of all human cultures and conscious thought.

But does mathematics exist independently of the mind? There are several different "orthodox" opinions about this. Most professional mathematicians and scientists, if pressed, would probably argue that mathematics is already "out there" in some (ill-defined) form waiting to be discovered. Accordingly, the only contribution of our minds is to bring mathematics to light. This viewpoint may appear the most simple and aesthetically attractive, but it runs into problems as soon as we ask: *Where* does mathematics reside if not in the mind? Where do mathematical concepts like "points," "the square root of two," and "pi" go when we are not thinking about them? The answer cannot be in the space and time of the physical universe because we can conceive, for example, of spatial geometries in many dimensions that (as far as we know) have no place in reality. The "discovery" argument also ignores at its peril the filtering and manipulating influence of cognition—the inescapable fact that everything we know and experience is a fabrication of the conscious mind (even if this fabrication is based on external effects). The inner world of our consciousness is the only one we have direct access to and, therefore, the only one in which we can have any real confidence. We should at least be skeptical of the notion of an abstract mathematical cosmos that allows us to interact with it, even though our cognition is limited to things in the universe of space and time.

The antithetical position is that mathematical structures and patterns are purely a human invention. Supporters of this idea are divided between two principal schools of thought. The first is that we choose our mathematics so as to provide a reasonably accurate model of the physical world; the second is that mathematics is no more than an elaborate game of our devising, with only occasional and fortuitous applications to the real world. Both fail to explain adequately why, if mathematics is simply man-made, different mathematicians in different cultures at different times have invariably come up with the same basic rules and concepts.

To these conventional viewpoints we can now tentatively add another. It is that mathematics, in common with all of our abstractions, depends for its existence on a synergy between mind and what lies beyond mind. In other words, mathematics does not reside exclusively "out there," or exclusively in the workings of the rational brain, *but requires both simultaneously.* In a sense, this proposal takes the middle ground in the debate over whether mathematics is discovered or invented, and so avoids the main weaknesses of the orthodox views. Yet it goes far beyond merely trying to reconcile some philosophical differences. It is, in fact, an interpretation of mathematics that casts man—or intelligence—in a profoundly creative role. At the very least, it implies that unless the world is governed by a specific system of logic then conscious beings cannot evolve. And it may go well beyond that. It may mean, if we choose to pursue the anthropocentric line to its extreme, that the logic of the world cannot exist without the intermediation of our minds.

Of course, metaphysical problems such as the nature of mathematics are not amenable to simple "right" or "wrong" solutions. Everyone—expert or not—is entitled to their own opinions and beliefs. It is only when speculation gives way to actual experiment that we can begin to test certain hypotheses about the universe in which we live. These experiments, as we shall see, reveal a remarkably close correlation between mathematics, conscious observation, and the physical world.

6

Mathematics and Reality

Mathematics began simply as a tool of practical reckoning. But so powerful did it prove to be, and such power did it bestow upon those who wielded it, that it became a focal point of intellectual life. One man, in particular, found an early fascination with it. Born around 580 B.C., he was Pythagoras of Samos—one of the first mathematical geniuses of ancient Greece.

To call Pythagoras a mere mathematician, though, would be to understate his influence. Later in life he was revered almost as a religious leader, and certainly as a philosophical seer. More than anyone before him, Pythagoras could look penetratingly behind the scenes of nature's play. His was the first mind to catch sight of the mysterious goings-on below the surface of things. He spoke of a unity in nature's variety, an underlying harmony. And that harmony he made visible, for Pythagoras found a basic relationship between musical harmony and numbers.

Vibrating as a whole, a stretched string will produce its lowest note—the fundamental. Divide the string into two equal parts and, as Pythagoras showed, the new note is harmonious with the original. Divide the string into three, four, or any other whole number of equal parts and its vibrations will be in harmony with the fundamental. That is, they sound pleasing to the Western ear. Divided in any other way, however, and the result is a discord. Not surprisingly, Pythagoras believed he had plucked from nature some deep insight. It was a belief strengthened by his discovery of other satisfying mathematical relationships in the field of geometry.

Unfortunately, Pythagoras and his many disciples went a little too far. They became obsessed with seeking out the divine harmony of the world—the "cosmos," as they were the first to call it—in order that they might become one with it. In fact, they were so blinded by this mystical imperative that they made the blunder of assuming that numbers themselves were all-significant. As Aristotle later commented: "They held, for instance, that ten is a perfect number. On this view, they asserted that there must be ten heavenly bodies; and as only nine were visible they invented the 'counter earth' to make a tenth." Of course, that sort of cavalier approach strays far beyond the bounds of good science (not to mention good mysticism). And yet notwithstanding the cultic quasi-religion he helped spawn, Pythagoras made immense positive contributions. His mathematical discoveries, we can appreciate today, began a chain of philosophical inquiry that led, ultimately, to our modern world view.

The next link in that chain was forged over a century later by another Greek colossus, Plato. Again, it can be argued that Plato's methodology—if not his ideas—fell well short of today's rigorous scientific standards. And there is no doubt that he chided those who would try to understand the world by making observations of it.

To Plato, the truth about the universe was contained exclusively in what he called "Forms," these being the perfect blueprints of all that we see. He considered the world of the senses—the world as it is actually experienced—to be only a flawed copy, or shadow, of an immaculate, eternal cosmos of the intellect. Observed happenings, therefore, were less fundamental to him than the unchanging aspects of the Forms that governed them. Indeed, the whole thrust of Plato's philosophy was to expunge the process of change and the notion of time from the description of things.

It was Plato who first fully grasped the notion, implicit in modern science, that there may be a complete theoretical counterpart to the world as we see it. Where he went astray, however, was to insist that his Forms could be deduced entirely by intellectual effort. The weakness of that assumption is clear: Plato would not have been led to the idea of Forms in the first place if he had not experienced—as we all do—the world as it reaches our senses. Fortunately, his stubborn denial of the value of observation was soon to be vigorously challenged—and by one of his own pupils.

Beyond doubt, Aristotle was one of the most influential thinkers who ever lived. He started out, as a student of Plato, by composing dialogues in his master's style. But after setting up his own school, the Lyceum, it soon became clear that he had no time either for the idealism of the Platonists or the mysticism of the Pythagoreans. Aristotle was a pragmatist, an ardent realist, whose interests ranged over all aspects of the familiar world. His passion was for collecting and classifying every scrap of observational data available to him, whether it had to do with plants, animals, rocks, or the objects in the sky. From this, we might easily conclude that he was among the first true scientists. But that would be a mistake. Aristotle was not a scientist in the modern sense because he failed to try to seek any relationships between the data he so meticulously gathered. Moreover, he never attempted to manipulate nature artificially in order to test his ideas. He collected, dissected, and organized, with scrupulous attention to detail, but he never experimented. And the reason for this stemmed from his underlying view about the nature of the universe and the destiny of every material object it contained.

Unlike Plato, Aristotle gave precedence to the world as we experience it—the world of observation. He rejected the idea that perfect, eternal Forms represented the only true reality. Yet—and this was central to his philosophy—he retained a modified notion of Form as one of two basic aspects of all material things. The other aspect he identified as "Substance." In Aristotle's view, Substance was the stuff we touch and handle, while Form was its specific realization. As distinct from Plato's Ideal Forms, Aristotle's Forms were an intrinsic part of things and could change over time. How they changed was determined by four distinct types of Causes: a Material Cause, a Formal Cause, an Efficient Cause, and, most crucially, a Final Cause. These causes, and their resultant effects, Aristotle maintained, were inherent properties rather than (as we would consider them now) relationships between events.

Suppose, for instance, that we carve a statue from a block of marble. The Substance is the marble, and will remain so, but the Form—the configuration that the marble takes—clearly alters as we transform the rough-hewn stone into a finished work of art. The evolution of the Form, in Aristotle's view, is decided entirely by the intrinsic Causes. What the thing is made of—in this case, marble—is determined by the Material Cause. The specific design of the object stems from the Formal Cause. The agent—namely, ourselves, the sculptors—that produces the object by embodying the Form in some material substance is the Efficient Cause. And lastly, and most decisively in Aristotle's scheme, is the purpose for which the thing exists. This is the Final Cause. Every Form, in Aristotle's universe, moves unerringly toward some ultimate, idealized state as if magnetically drawn to it. In fact, Aristotelean philosophy is unreservedly teleological, embracing in full the notion that what is happening now is compelled by some inevitable destiny. Moreover, the end of all things, as Aristotle saw it, was not just what happens last, but a state of absolute and flawless harmony. All motion, all activity in the universe, he considered to be leading inexorably to this perfect final state.

* * *

The great natural philosophers of ancient Greece may seem to suffer from an excess of intellectual arrogance. They genuinely believed they could solve the deepest of cosmic mysteries through pure thought (abetted in Aristotle's case by a mass of taxonomic data). Yet, surprisingly, given this self-imposed restriction, they arrived at some conclusions that even today appear quite reasonable. In a sense, Plato and Aristotle stood as advocates for the two principal ways we have of looking at the world. They personalized the underlying dichotomy of the universe.

Plato was essentially a holist. He saw the real truth and substance of the universe as being the timeless Forms that lie behind the impurity of perceptions. To him, all that mattered was the eternal, unifying logic of the cosmos.

Aristotle, on the other hand, was a reductionist. He gave precedence to the discrete objects and events of our experience, insisting that these are the primary reality.

Later on, it will become clear that there are some deep resonances between both the Platonic and Aristotelean philosophies and certain interpretations of our modern scientific world view. In particular, the teleological argument of Aristotle, as we shall see, finds a curious echo in a radically new idea known as the Participatory Anthropic Principle.

On the debit side, there is no doubt that the Greeks unwittingly did a lot to hold up the progress of science. This is especially true of Aristotle. His influence on future generations across Europe was simply too powerful, too pervasive. For two thousand years, the cosmos of Plato and Aristotle went virtually unchallenged. To give an example, Aristotle formulated some "laws of motion" based upon his doctrine of Final Causation. One of the implications of these is that heavier objects should fall faster than lighter ones. In some cases, of course, this seems to be true—a stone does fall faster through the air than a feather. No one thought of challenging Aristotle's word. No one, apparently, considered putting his "laws" to the test. The dictates of the Greeks became like an inviolate religion. Indeed, much of ancient natural philosophy became, if not absorbed, then at least condoned and approved, by orthodox religions in the West.

In the end, however, it was the Judaic, Christian, and Moslem traditions, which (despite being so dogmatic) actually helped nurture the open-minded scientific approach. The belief in a rational deity, who consciously designed the cosmos, encouraged the idea that there must be a coherency and a natural order to things—from which it followed that by observing the world it ought to be possible to elucidate this underlying order. That prospect was the driving force behind the Renaissance intellect and its personification in men like Kepler, Galileo, and Newton. But though these scientific pioneers took their cue from theology (as well as from the Greeks), they were soon to find themselves increasingly out of step, and out of sympathy, with the Church's doctrinaire teachings.

Central in all this was Galileo, for it was with him that modern experimental science really begins. Galileo's genius was to see that what appears to be a very complex process may actually be a very simple one in disguise. For example, from his investigations of falling objects he realized that a great many influences were at work: the shape of an object, the distribution of its mass and internal motion, the wind speed, the density of air, and so on. But he realized also that these were incidental factors; that they were complications superimposed on a plain underlying truth. So he began systematically to eliminate or control the distractions, by using bodies of regular shape and by ensuring the air was still. At the same time, he made the quantity he was interested in—the time taken to fall a certain distance—easier to measure by rolling objects down gentle inclines (instead of just letting them drop). Effectively, he did in practice what for ages man had been trying to do with his mind alone. He excised some small part of the world from its surroundings and gave it the freedom to behave as simply as possible. His experiments demonstrated that, in fact, Aristotle had been wrong. All objects, whatever their weight, fall to the ground at the same rate. Furthermore, the time taken to fall a given distance from rest, he found, is exactly proportional to the square root of that distance. For his troubles, Galileo had uncovered a basic natural law. And that law, it turned out, was mathematical.

In the year that Galileo died, 1642, a new champion of science and of mathematics was born. He was Sir Isaac Newton, a man whose searchlight intelligence lit up the physics not only of the past and the present, but also of the future. Optics, dynamics, gravity—each in turn yielded to his brilliant scrutiny. But no achievement of Newton's seemed greater than his conquest of time. Given the exact present state of a physical system, Newton propounded, the state at any given point in the future could be worked out.

But he could not do that with the mathematics at hand. And for a simple reason: there was no provision in it for dealing with continuous change. Newton had to extend mathematics so that it could take account of the flux of time. In fact, he called his new technique "fluxions," though it is by a less picturesque name (due to Leibniz) that we know it today: the differential calculus.

The key point about the discovery of the calculus is that it shifts mathematical thought from the static to the dynamic, from a series of still frames to a smoothly flowing, dynamic process. That is important because, as Newton and others quickly found out, most of the laws of nature can be expressed mathematically as differential equations. In other words, they involve quantities whose rate of change depends on the rate of change of one or more other quantities. A differential equation is like a door leading from the present to the future. Open that door and, knowing the present state of a physical system, you can find out what all of its future states will be.

For nearly two centuries after Newton's death, his central belief held firm—that, in essence, the cosmos was a clockwork mechanism. Not only was the future held absolutely within the present, but he could calculate that future to any desired degree of accuracy. It was Pierre-Simon Laplace who put the Newtonian claim most forcefully. Find out the positions and speeds of all the bits of matter that exist, he pointed out, and by solving the differential equations governing their motion it would be possible to know the precise state of the universe at any future time. That would include an exact knowledge of all the actions that everyone would ever make and all the thoughts and decisions that would ever enter anyone's head. It seemed as if the program of theoretical physics would soon be accomplished—and with it the death of free will.

Except that nature was not ready to have her secrets so easily wrenched away. Early in the twentieth century, a scientific bombshell exploded that reduced the Newtonian clockwork cosmos to ruins. Effectively, it put an end to the hope (or the despair) that you could ever accurately know the future. But even as mankind came to grips with this devastating truth—that it could never be totally certain what tomorrow would bring—it learned that it might have a far greater power. Man, it turned out, was not a mere spectator to some vast cosmic clockwork, nor a trivial cog in a machine whose every action was preordained. Incredibly, it transpired that man might be an active and vital player in determining what was real.

7

Quantum Sorcery

We are conscious of only a tiny portion of the phantom energies that fall upon us. And that is fortunate, because if we had senses that could discriminate every individual light ray, every radiation beyond light (including radio waves, ultraviolet, infrared, and X-rays), and every sound wave of every frequency, our brains would be hopelessly overloaded with information. Moreover, from a survival standpoint, most of it would be useless information. That is the very reason we have not evolved to be able to detect it. What is the good of having eyes (or aerials) sensitive to radio waves if nothing that is important to you shines brightly in that region of the spectrum?

It is the same with our experience of matter. We see, and are otherwise naturally equipped to sense, matter on a scale that is relevant to our survival. We can resolve details of something close up, for example, down to about a hundredth of an inch across—and even that is better than we need for most practical purposes. For a hominid to admit into its consciousness all the goings-on at the microscopic level would be a lethal distraction.

Until recently, then, ours was a highly parochial and personalized cosmos. But now, through our technology, we have rapidly, massively, extended our senses, and therefore our consciousness, of what is happening outside the familiar world of our biological eyes and ears. With radio telescopes we really can now see, or listen to, the universe at very long wavelengths. We have built X-ray detectors and instruments sensitive to other, previously invisible radiations. And we have devised equipment that can tease apart the inner structure of matter and energy—the grain within the grain.

Is it so surprising that what we find in these unfamiliar realms is vastly different from anything we have experienced before? Surely not. We would not go to an unexplored land and expect the natives there to speak our language or to share our culture. By the same token, it is unreasonable to presume that regions of nature inaccessible to direct human sensation should follow the same pattern as those we are accustomed to.

In probing ever deeper the structure of matter and energy, we must be ready for surprises—even shocks. And they begin early on in the descent. Look through a microscope at a thin section of animal tissue and you no longer see a bit of animal, a small piece of flesh. You see cells;

and there is no way to predict that extraordinary internal architecture from what your eyes alone reveal. Use a more powerful microscope and you can penetrate the individual structures within the cell—the cell nucleus, for example—and look upon yet another hidden stratum of organization.

What does that view tell us? Most obviously that nature's patterns change dramatically as we move to smaller and smaller scales. The view at a scale length of, say, six feet, which would include the whole human body, is in sharp contrast to that at one thousandth of an inch, embracing just a single cell. Of course, man was accustomed to that sort of wholesale shift of perspective long before the advent of powerful scientific tools. Rome appears very different from the top of one of its hills than it does from the gates of the Colosseum. The broad sweep of the savannah from horizon to horizon possesses a dynamic and a type of organization that bear no resemblance to those of a single blade of grass. The phenomenon is universal. Level upon level, nature is built up; each level is in continuous interaction with those immediately above and below it, but each possesses its own distinctive character and architectural style.

And yet we must try to look beyond these idiosyncrasies of nature. That has really been the task of science all along: to see through the many varied ways in which the cosmos presents itself, to the root principles. What lies at the most basic level of reality?

Take a cell. Does it behave in the same way as the human being of which it is an essential part? No, because a human being can talk and imagine and perceive a harmony in nature, while an individual cell can do none of these things. The human—a system of one hundred trillion cooperating cells—can perform in ways that a single cell obviously cannot. But that tells us nothing about fundamental laws. It tells us simply that as the degree of complexity of a system increases, properties may come into being that were not there before—self-awareness, humor, the ability to think as we are doing now. These are unpredictable, emergent properties. You could never show beforehand that one hundred trillion cells, brought together in such-and-such a way, would give rise, say, to poetry. Poets and their works are emergents not contained within nature's more basic laws, and yet, crucially, not in violation of them either. Man has a mind, but the brain that subtends the mind is subject to the same law of gravity as every other object in the universe. The firing of neurons that on one level gives rise to a poetic or scientific thought is, simultaneously, operating in perfect accord with the underlying principles of electromagnetism.

What, then, are the finest constituents of the real world? Not cells, not molecules, not even the atoms of which molecules are made; because you can pry apart an atom and liberate the smaller components within—the protons and neutrons inside the nucleus at the heart of the atom, and the electrons that orbit around the nucleus.

It begins to seem as if nature may be like an infinite Russian doll, revealing shell within shell, endlessly, as our probing instruments become ever more powerful. But, in fact, with the electron the journey inward comes to an abrupt end; the electron gives every appearance of being a finality— a truly "elementary" particle. The proton and the neutron, too, resist our best attempts to break them apart. Yet, in their case, we can discern, indirectly, one last level of structure. Within each proton and neutron, we have good reason to believe, is a triplet of tinier, but permanently inseparable, building blocks known as quarks.

The entire stable material world, if our current notions are right, is comprised of these two fundamental entities: the electron and the quark. But merely by naming them we have not explained them. We need to ask what an electron and a quark really are. We need to inquire what the universe is like at this fantastically small scale.

All of our language is founded upon personal experience; it could not be otherwise. But at the frontiers of the atom (and, indeed, well before that) we are forced to abandon the world of the eye and the ear, and enter a strange, unfamiliar land. As one of the great pioneers of subatomic physics, the Dane Niels Bohr, remarked: "When it comes to the atom, language can be used only as in poetry. The poet, too, is not nearly so concerned with describing facts as with creating images."

One obvious possibility is to try to imagine the electron or the quark as being just a tiny speck of everyday matter, like a little bead or a fantastically small billiard ball. But, at the subatomic scale, reality revolts against such a comforting, commonsense notion. Yes, at times, an electron can seem to masquerade as a particle—a solid, material dot. Yet, at other times, it breaks out of the metaphor, and shatters the illusion that we have managed to visualize it. Because at times an electron, in company with every other type of subatomic "particle," acts as if it were smeared out over space in the manner of a wave.

Suddenly, we find ourselves grappling with two discordant images, one of a particle, one of a wave, and somehow trying to make the two fit the behavior of matter inside the atom. But that is only the beginning of our conceptual problems. It turns out that light—a form of energy, not of matter—also shares in this schizophrenia. A light ray has both a wavelike and a particlelike face. That is, it can appear either to be spread out in space or focused at a single point, as a particle called a photon. How can we bring this bizarre dualism of subatomic matter and energy out into the open?

There is an experiment that has been described a thousand times but is no less infuriatingly delightful and profound for all that repetition. It involves simply a source of photons, a screen with two parallel slits, and a photographic film. We set up the light source in front of the slits, switch it on, and observe the effect on the film placed on the other side of the slits.

What we see is an interference pattern: a series of alternating bright and dark bands. Immediately, that speaks to us of wave action. Two waves, one rippling out from each slit, have combined, and where, on the film, two wave crests or two wave troughs have coincided they have doubled up and produced brightness; where a crest and trough have come together they have canceled each other out and left the film dark.

So far, there is no mystery. If light can behave as waves you would expect it to make an interference pattern, just as water waves and sound waves will do. No, the real puzzle, the puzzle that like a Zen koan puts us in touch with the infinite, becomes apparent as we begin to dim the light source. Gradually, we turn down the intensity of light until we can see individual spots building up a pointillistic picture as the trickle of photons meets the film. It is the interference pattern again, emerging slowly but steadily. And as the pattern builds, so does the mystery:

how is it that the photons can behave as particles when they strike the film, but as waves while they are en route from the light source? For that is what the results imply. There is the interference pattern, which is indisputably a wave phenomenon. And there are the discrete points on the film, which are just as plainly the impact marks of particles. How can a photon be a wave one instant and a particle the next?

To bring the problem into still sharper relief, we can turn down the light source even further. In fact, we can turn it down so that it gives off only one photon at a time. As each photon hits the film it becomes like a little meteor, burning up in a speck of brilliance. At this instant of collision, when simultaneously it reveals its whereabouts and dies, it is a particle. Yet, astonishingly, the pattern that accumulates, dot by solitary dot, is an interference pattern as before.

What does it mean? Apparently, that some property connected with each photon, as it comes to the screen, passes through, not one or the other of the slits, *but through both*. Each photon interferes with itself! And it makes no difference whether we use photons, or electrons, protons, or any other kind of subatomic particle in the experiment; the result is the same.

In the 1920s, when the brightest minds of Europe were first turned on this and related quantum puzzles, at the University of Göttingen in Germany, they would say that on Mondays, Wednesdays, and Fridays an electron behaved like a particle; and on Tuesdays, Thursdays, and Saturdays it behaved like a wave. On the Sabbath, added the physicist Banesh Hoffmann, the theorists simply prayed.

The problem is, we have to square two quite different descriptions of a single entity without any hope of a verbal reconciliation. We lack the words and mental imagery to explain really what is going on beyond the bounds of our experience. That is why all our metaphors and analogies to do with ultrasmall phenomena must ultimately fail. Snatched from the Brobdingnagian world of the familiar, they are hopelessly inapt in the domain of the subatomic. But if we are left speechless, we are far from being rendered powerless to understand the quantum world. Potentially, we retain the most potent descriptive tool of all: mathematics. And curiously enough, it turned out that part of the mathematics needed to successfully confront the bizarre wave-particle duality of matter and energy—and much else besides—was already well known, long before men had any inkling that nature behaved in such an outrageous way.

Some of the mathematics used in the formulation of what has become known as quantum mechanics was not previously thought to have any practical application. It was a mental abstraction—an interesting game for those who could play it, like chess. Except that, in the first quarter of this century, it became far more than a game. Following man's breakthrough into the subatomic cosmos, it was found to choreograph in exquisite detail the dance of matter on this smallest of scales.

Actually, there are two different mathematical ways to describe the subatomic world, though these have been shown to be exactly equivalent. There is a scheme known as matrix mechanics, whose chief architect was the German Werner Heisenberg. And there is another, which on paper looks as different from matrix mechanics as, say,

Arabic does from Chinese, developed by his compatriot Erwin Schrödinger. The latter is called wave mechanics, because the central object in Schrödinger's formulation is a thing called the wave function. It is this wave function that expresses most graphically what we failed to capture with conventional words: the complementary make-up of the microcosmos.

The wave function is a quantity associated with every particle. It simply enshrines mathematically the observational fact that electrons, photons, and the like can behave like waves. A crucial question is: What is the nature of these waves?

At first, Schrödinger himself suggested that the wave function of an electron, for example, might be related to how the electron's charge became smeared out as the particle moved through space. According to this idea, the charge would have to consolidate instantaneously at a single point as soon as the electron interacted with a measuring device. However, such instantaneous motion would violate Einstein's special theory of relativity, which forbids matter or energy to travel faster than the speed of light. An alternative proposal was therefore put forward by the German Max Born. He interpreted the quantum wave associated with a particle, not as a physical effect, but as a wave of probability. Regarded in this way, the wave function of an electron gives the odds of that electron's turning up as a "real" particle, at a certain place, if we make an attempt to look for it. Born's idea, advanced in 1926, was quickly adopted and extended by Bohr to give the so-called Copenhagen interpretation. It is this extraordinary reading of the quantum mechanical formalism that the majority of physicists accept today.

Most astonishingly, in its original form, the Copenhagen interpretation denies that there is any physical reality independent of Man the Observer. Our decision to observe, it insists, is crucial, because unless and until we carry out an observation an entity like an electron *cannot be said to exist at any place or in any discrete material form at all*. In between our attempts to pin it down, it has no existence outside of the probabilistic description given by the wave function. It is not simply that we lack the technical means to keep track of it, instant by instant, or that our theories are not yet precise enough to work out just where the electron is. The fact is, there are no electrons or protons or photons "out there" until we go and look for them. In Bohr's own words: "An independent reality in the ordinary sense can neither be ascribed to the phenomena nor to the agencies of observations."

Not to overstate it, the Copenhagen interpretation as posited by Bohr demands a radical reappraisal of our world view. It insists that when a measurement or an observation takes place—for example, when a spot appears on the recording surface of the two-slit experiment—we should not think of this as bringing to our attention the whereabouts of a previously invisible particle. Prior to the observation of the spot, *there was no particle*. It is not correct to say that a particle "collided with the film and produced a visible trace." If we accept the Copenhagen interpretation, the truth is that the observing apparatus interacted irreversibly with the wave function of the electron, causing it to telescope down to a single point in space and time. At that moment of interaction, the wave function was forced to abandon its probabilistic description of many possible alternative states for the electron (all of which are equally "real" or "unreal") and select just one—the one that is actually observed.

Nevertheless, it is tempting to suppose that you could use the results of an observation to work backward and say, "This is where, and in what condition, the particle must have been had the observation *not* taken place." Einstein himself tried repeatedly to devise experiments that would show such hindsight was possible. But in the end, even his great Luddite attempts failed to hinder the quantum revolution.

The point is—and, again, this is implicit in the mathematics of quantum mechanics—not only does the observation process collapse the wave function, but it does so *in ways that cause a fundamental, unknowable disturbance in the system*. That disturbance, when something new and unforeseeable is made to happen at the ground level of reality, can be regarded as a fundamentally creative process. It is the very essence of what true creativity must be: not building upon anything that is already there, but spontaneously bringing into being that which did not exist, in any meaningful form, before. In a sense, this is genesis at work, here and now: the making of the real from the unreal, the breathing of fire into the equations that underpin the world.

It is remarkable that such a deep metaphysical insight should spring from the heart of what is now a mainstream scientific theory. But from what we have already seen of the relationship between human consciousness and the external world, we should not have been unprepared for it. Indeed, it comes as a powerful corroboration of what we have been suggesting all along: that the conscious mind is crucially involved in establishing what is real. That which reaches our senses is, at best, a confusion of phantasmal energies—not sights, or sounds, or any of the coherent qualities that we project outward onto the physical world. The universe, as we know it, is built and experienced entirely within our heads, and until that mental construction takes place, reality must wait in the wings.

What quantum mechanics does is offer the possibility of putting these iconoclastic ideas onto a sound theoretical framework, girding them with mathematics, and enabling them to make empirical predictions. The wave function of a particle—a purely mathematical thing—is the only reality there is, until an observation takes place. The wave function evolves in time. That is, it changes, instant by instant, to give the new probabilities for finding the particle at any given point in space. There is nothing fuzzy about this evolution. Often, the mistake is made of thinking that nature unfolds haphazardly at the subatomic scale. But, on the contrary, the evolution of the wave function is clinically precise and deterministic. Once it has been specified, its entire future behavior becomes known—*providing there is no attempt at a measurement.* As soon as we ask where a particle is (for instance, by putting a photographic film in the way), we change the rules of the game. At the moment of observation, the particle, or the effect we associate with the entity we call a particle, is seen to be at a definite place. Perceived reality, the Copenhagen interpretation implies, becomes crystallized. At that same moment, the wave function has to undergo a sudden and spectacular transformation. Instead of being spread throughout space, giving the likelihood of finding the particle anywhere, it has to abruptly "collapse" so that it agrees with the actual measurement. This collapse of the wave function (known technically as the *state vector reduction*) is the direct outcome of the intrusion of our macroscopic measuring instruments into the affairs of the submicroscopic world.

Clearly, we need to look at what these conclusions might imply on a wider scale. In particular, we need to ask whether the observer-induced creation described by Bohr is a potential mechanism for bringing, not just subatomic particles, but large portions of the universe into being—including ourselves.

8

God's Eye

With each observation that causes the wave function to collapse, the Copenhagen interpretation implies, a small part of the universe is made real. But what constitutes a valid quantum observation or measurement? The only mathematical requirement, in turns out, was first shown by the brilliant Hungarian-American mathematician John von Neumann. It is simply that the measuring apparatus must be *excluded from the wave function of the system being observed*. Unless this is done, the wave function cannot be made to collapse.

Now, there are two complications in establishing a

clear distinction between what is observing and what is being observed. The first is that everything in nature is actually connected or implicated with everything else. That is bound to be the case in a universe in which forces such as gravity and electromagnetism form a permanent glue, albeit a tenuous one, even between subatomic particles billions of light-years apart. The other problem, in distinguishing the observer from the observed, is knowing exactly where to draw the line. That is, what constitutes the minimum arrangement needed to collapse a wave function? Does it really have to involve a sentient observer (as Bohr himself believed), or would a Geiger counter or some other kind of recording instrument be able to do the job on its own? And if an intelligent observer was needed, could it be an ape or a young child instead of a "fully conscious" adult? Would perhaps a smart computer do?

Von Neumann's contribution was to show that the measuring apparatus, whatever it might consist of, has to be on a different physical level from the thing being measured. This means there must inevitably be a downward chain of causation. In other words, something at a higher level (such as a Geiger counter) must induce a fundamental change in the state of a lower entity (say, the wave function of an electron). It is absolutely crucial, Von Neumann proved, to achieve that demarcation between the macroscopic and the microscopic. You cannot collapse a wave function with an observing device that is itself considered to be just a collection of quantum particles. Such a device would form part of a larger wave function that embraced the quantum state of the system it was supposed to be measuring—and it is impossible for a wave function to bring about its own collapse. The very meaning of this type of measurement refers to our drawing a distinction between the microscopic level of subatomic particles and the macroscopic level of complex, "classical" pieces of apparatus, in which irreversible changes take place and traces are recorded.

And so we return to the key question of what level of complexity is needed to successfully trigger the collapse of a wave function. Put another way: how complex must reality be on the large scale in order to participate in making things real at the elementary scale?

In seeking an answer, it may be helpful to view the process of downward causation in terms of information—and, specifically, in terms of the way information is handled by computers. The wave function is analogous to software; it holds all the instructions governing the present and future state of the quantum system. By contrast, the particle can be thought of as hardware. So the wave-particle duality of matter has a close parallel in the software-hardware duality of a computer. Just as with a computer there are two complementary descriptions of the same set of events, one in terms of the program and another in terms of the state of the electrical circuitry, so an entity such as an electron can be described in two complementary (and mutually exclusive) ways. The difference—the vital difference—is that, in the case of the quantum measurement, what is seemingly a closed system (particle plus measuring apparatus plus observer) evolves in such a way that there is a change in the software that brings with it a change in the hardware. That is, the wave function—the quantum program—jumps discontinuously to a new state through the intervention of the measurement, while, at the same time, a particle is momentarily created whose future behavior is unknowably different from what it would otherwise have been.

How is it that the mere act of observation can make that happen? Perhaps we could simply ignore the issue if it were not for the fact that, in principle, the effect of a single subatomic event could be amplified until it had dramatic large-scale consequences. The man who developed the mathematics of wave mechanics made the point startlingly clear in a thought experiment that has become known as Schrödinger's cat.

It is quite simple: a luckless cat is put into a room in which there is a flask of poisonous gas. Poised over the flask is a hammer. By way of a detection and amplification system, the hammer will only fall and break the flask, whose freed contents will kill the cat, if a single radioactive nucleus stored nearby splits up. The behavior of the nucleus is described by a wave function that is the sum of two separate wave functions—one covering the possibility that the nucleus does spontaneously decay; the other covering the possibility that it does not. Until an observation is made, the overall wave function remains a continuously evolving mixture of the two possible states. In this unobserved condition, the nucleus as a particle with definite properties has no reality, and it is meaningless, even in principle, to ask whether the decay has or has not happened. The only reality is the statistical description of the combined wave function.

That is all very well. But the fate of the nucleus decides the fate of the cat. So what we really have is a situation in which there is a superposition of two wave functions covering both possibilities for the cat: a dead-cat wave function plus a live-cat wave function. In other words, the whole content of the room is subject to a wave function that is a superposition of the living- and the dead-cat states. The only way to force a decision, to collapse the global wave function, is to make an observation.

Again, what we have to understand is that, according to quantum mechanics, before we open the room the cat is neither dead nor alive. In fact, because we are treating it as part of a large quantum system, it has no material reality at all until a higher system—a macroscopic measuring system—looks in upon it.

Of course, it seems utterly bizarre. And you may object, pointing out that, while we may be uncertain about the cat's fate until we check it out, surely the cat knows. But no, that apparently reasonable, commonsense loophole is not available. If something is described by a wave function, as the cat in this case is, its exact condition—its very existence—is devoid of meaning until a valid quantum mechanical observation is made.

We could replace the cat in the experiment by a human being, or even by a roomful of physics professors, and it would make no difference. The reality of what had happened inside the room would not only be unknown, it would not exist until an observation took place that caused the collapse of the combined wave function.

Yet surely there is a limit to how far we can push this. Imagine, for instance, that the entire human population with one exception—yourself, say—was crowded into a great building where an enormous mallet hung in judgment over a monstrous globe of gas. The arrangement is otherwise the same as before. Are we really claiming that the fate of the human race is only decided at the instant you poke your head around the door? And if we sheepishly nod "yes," would we likewise agree if the person outside was a child? How old do you have to be to be a qualified quantum mechanical observer? Eighteen? Ten? Two? And do you even have to be human? Would the experiment be valid with a chimpanzee or a monkey or a mouse looking in?

We have come so far, why not push Schrödinger's strange test to the extreme? Suppose that all the galaxies throughout space have been planted with mines and that all the mines can either be detonated or not at the whim of a solitary radioactive nucleus. Just one person . . . no, let it be a cat . . . just one space-suited cat is left floating free in the intergalactic void. For one hour the cat dozes, and then opens its eyes. Is the experiment still valid? Does the cat cause the wave function of the rest of the universe to collapse to one of its two possible states because, ultimately, that wave function is coupled to the wave function of a single particle?

These extensions to Schrödinger's thought experiment, though apparently outrageous and whimsical, actually pose very serious questions. As the physicist John Wheeler puts it: "How is it possible that mere information [the wave function "software"] should in some cases modify the real state [the "hardware"] of macroscopic things?" In other words, how is it that just an interrogation or measurement of the wave function of a subatomic system is able to affect not only the fate of a single particle but potentially the physical state of something very large indeed? Any measurement of a quantum system, as Bohr originally asserted, calls for an irreversible amplification, which results in a macroscopic record or trace. But Wheeler (a one-time colleague of Bohr's) takes the requirement further. The record, he insists, must be meaningful. That is, it must be given meaning through conscious observation.

And so we are back once more with the mind—and, in particular, the human mind. Because it is only man, as far as we know, that can perceive the subtle interconnections of nature, the threads of meaning that link concepts and events experienced in the brain. According to this view, then, a measuring device (a photographic film, a Geiger counter) is not sufficient on its own to cause wave function collapse; nor is it enough for the device to be monitored by a brain that cannot understand the significance of what it sees—such as that of a cat or an ape or a small child. Indeed, it may not even be sufficient to have a single, fully developed human mind at the end of the measurement process. Wheeler appeals, instead, to a "community of investigators" as being the minimum necessary to extract meaning from, say, the click of a Geiger counter or a spot on a screen. So, in this way, we can begin to trace a possible chain of causation from elementary particles through molecules and macroscopic objects to conscious beings and communicators and meaningful statements.

It is fair to point out that these issues are the subject of vigorous and lively debate. Wheeler's suggestions about the necessity of meaning and intersubject agreement, for instance, are essentially matters of opinion (as he is the first to admit). In fact, devising experiments to test the various theories about the mechanism of wave-function collapse is extremely difficult, and there are certainly no conclusive results yet. For this reason many scientists avoid becoming involved, on a professional level, with philosophical discussions about quantum theory, preferring instead to concentrate upon its practical applications. They would probably concur with Stephen Hawking's remark: "When I hear of Schrödinger's cat, I reach for my gun!"

Among scientists who are prepared to speculate openly, there is a healthy diversity of views. At one extreme are those who insist that the whole business of wave-function collapse and quantum indeterminism is a sham. According to this school of thought, particles and their properties really do exist independently of observation, and the only reason quantum theory is statistical is that it excludes some (as yet unidentified) classical variables, which actually govern the behavior of atomic systems. These hypothetical unknown factors left out of quantum mechanics are called "hidden variables" by the supporters of such a deterministic world view. Unfortunately for this idea, a number of experiments have been carried out over the past decade that uphold the predictions of quantum theory and rule out the existence of at least one major class of hidden variables. These experiments are generally regarded as providing strong support for the Copenhagen interpretation.

Yet even advocates of Bohr's stance differ markedly in their opinions. Some maintain that the true "observer" is just the apparatus that amplifies a subatomic event to macroscopic dimensions. If so, then Geiger counters and photomultiplier tubes are capable of creating reality themselves, without necessarily being supervised by human beings.

A very different and quite extraordinary theory has been put forward by Hugh Everett III, a one-time graduate student of John Wheeler's at Princeton University. Everett sidesteps the entire issue about when and where observer-created reality takes place by claiming that wave functions in fact *never* collapse. He envisages a situation in which every time a quantum interaction takes place—whether it is observed or not—the wave function simply splits apart to take account of all the possible outcomes. Whereas Bohr's interpretation leads to a breakdown of determinism because it only allows us to predict the *probability* of what we eventually observe, Everett's proposal goes out of its way to preserve determinism: each of the possible states, it maintains, really occurs. This so-called "Many Worlds Interpretation" predicts that every time an event takes place at the subatomic level—even if it is as trivial as the collision between two elementary particles—the universe divides into a number of alternatives, which then pursue separate evolutionary paths. Since this happens an enormous number of times every second and has presumably been going on for billions of years, the result is truly mind-boggling. Somewhere, sometime, among all the host of continually-dividing "parallel" universes, everything that can happen does happen.

Considering the monumental suspension of disbelief called for by such a theory, it seems remarkable that so many leading quantum scientists are prepared to subscribe to it. One of the reasons they do so is that Everett's interpretation does away with any troubling references to "consciousness" or "observers" or subtle distinctions between quantum systems and measuring devices. Nevertheless, it has its own set of problems—quite apart from its spectacular lack of economy. Most notably, it fails to give any account of why our conscious minds, despite supposedly splitting again and again with every quantum interaction within our brains, appear to follow only one track through the bewildering maze of Everett branches.

Whether we like it or not, consciousness has a persistent habit of intruding into all our discussions about the nature of mathematics, physics, and reality as a whole. We cannot just step outside ourselves to discover what things would be like—assuming they still existed at all—if we were not here.

Both in mathematics and in quantum mechanics, then, there is a vigorous, ongoing debate concerning the extent to which our minds play an active role in deciding what is real. Quite clearly, these debates are interlinked. Quantum mechanical systems are described by a very precise set of mathematical equations. If our conscious minds somehow take part in selecting and subtending a particular system of mathematics, then inevitably they must also be implicated in deciding how particles behave in the subatomic world. The picture of reality that emerges from such an ideology proves to be extraordinarily self-reflexive.

Man and the brain evolved as part of the universe they subsequently looked out upon. Through the rational workings of the left hemisphere, man uncovers a system of symbolic reasoning, aspects of which seem to describe the workings of the physical universe with uncanny accuracy. The behavior of the cosmos, it emerges, is fundamentally mathematical. Yet those underlying mathematical rules by which reality conducts its affairs can only exist and be given meaning in the mind. There is no mathematics without mind. There are no mathematical laws without mind. But mind evolved from a cosmos that is—and apparently always has been—mathematical. Then again, man probes the universe at the subatomic level and finds that observation plays a crucial role in determining what is real—

that without conscious participation there is no reality in the quantum world. But we are all made of quantum particles. Without matter obeying the mathematical laws of nature there could have evolved no mind. Without mind to extricate meaning from the material world there could have been no mathematics. And without the intervention of mind to observe nature meaningfully at its finest scale, there could have been no matter.

At this point, the pragmatist can be excused for losing patience with such a fantastic state of affairs and reverting instead to the simpler, objective universe of Newton and the neoclassicists. But that would be a hasty move; we have not yet finished reviewing the evidence. In fact, the most compelling testimony that man may be much closer than we normally suspect to the creative hub of reality comes not from the submicroscopic world of quantum mechanics at all but from the universe on its very largest scale.

9

At the Edge of the Infinite

Ours is a universe conducive to the emergence of life and consciousness—a fact that is hardly surprising or puzzling. Only a universe in which the evolution of life is inevitable could, at some point, harbor beings capable of pondering how accommodating nature has been. The surprise and the mystery is not that the only universe we know happens to be inhabitable, but that the conditions required for the eventual evolution of life (of any kind we can reasonably imagine) are so specific. We owe our existence, it turns out, to a chain of astonishingly unlikely coincidences.

* * *

The litany of "cosmic coincidences" begins with the origin of the universe itself. The further we look out into space the more it becomes clear that we are living amid the debris of a stupendous explosion. The galaxies—or rather the clusters into which galaxies are grouped—are racing apart. Run their motions backward and it appears that all the matter there is must have been unleashed during a cataclysmic outburst, the Big Bang, between about fifteen and twenty billion years ago.

Two opposing factors control the overall dynamics and evolution of the universe: the kinetic energy of expansion and the gravitational binding energy of matter. There is no obvious reason why these cosmic tug-of-war contestants should be more or less balanced. But the strange fact is they are, and moreover it is a condition that seems vital for the development of life as we know it.

Life is predicated upon organized complexity. Complexity, in turn, can only come about if sufficiently large amounts of matter remain together long enough for interesting structures and substances to evolve. Had the Big Bang been just slightly more violent, all the matter in the universe would have scattered too quickly for it ever to draw itself locally together (by gravity) into clumps. On the other hand, if the Big Bang had been marginally less powerful it would have been followed by a prompt, global infall of matter, allowing too little time for anything as complicated as life to emerge. The universe teeters on a knife edge between eternal expansion and eventual collapse—a fine line upon which our existence depends.

The fact that space has three dimensions also appears fortunate from our point of view. Any fewer than three dimensions and the interconnections between the brain's neurons would have to be so labyrinthine as to be wholly unworkable. On the other hand, in a universe of four dimensions or more the Earth's orbit about the Sun would be unstable, hurling our planet into the frigid darkness of space.

Other coincidences bear upon the relative strengths of the four basic forces in nature: gravity, electromagnetism, and the so-called strong and weak forces. The latter two are unfamiliar in the everyday world because they operate only over subatomic distances. Nevertheless, these short-range interactions play a pivotal role in determining the nature and stability of ordinary matter.

Only about one percent of the material in the universe consists of elements more massive than hydrogen or helium. Yet, without the sprinkling of heavy elements, biological evolution of any conceivable kind would have been impossible. Their manufacture has taken place in the extremely hot interior of stars more massive than the Sun, and their dispersal has occurred when such stars blew apart in titanic explosions called supernovas. This process of stellar nucleosynthesis, however, is contingent upon both the strong and weak forces' having very particular strengths.

The fabrication of heavy elements inside stars happens through a stepwise series of fusion reactions. This begins when ordinary hydrogen nuclei, or protons, collide at high speed to form nuclei of deuterium. Further reactions, in which deuterium is an essential intermediary, result in the formation of helium, which, in turn, becomes the raw material for making heavier elements, such as carbon, oxygen, and sulfur. Without deuterium as a stepping-stone, none of the complex substances in our bodies could have been made; yet the very existence of deuterium would be impossible if the strong force were about 5 percent weaker than it actually is. A strengthening of the strong force by only about 2 percent, on the other hand, would allow bound states of two protons (diprotons) to form. In fact, these would proliferate inside stars and then quickly change into deuterium by radioactive decay. As a result, stellar cores would be so awash with deuterium—and the subsequent energy-releasing production of helium so enormously accelerated—that stars would almost certainly blow themselves to bits before they had a chance to manufacture the elements crucial to life.

The strength of the weak force must also lie within a very narrow band if the heavy elements produced in large stars are to be scattered into space during supernova explosions. These explosions happen immediately after a star with around twenty times as much mass as the Sun exhausts its useful nuclear fuel. The pressure of the outer layers of the star (previously resisted by outward-moving light and heat from the center) squeezes the inner regions so hard that electrons and protons are forced to merge with one another, forming neutrons. The resulting neutron-rich core is about as massive as the Sun but measures only a few miles across. This sudden shrinking of the core pulls the floor from under the outer parts of the star—the other nineteen solar masses or so—causing them to plummet downward at speeds of up to 28,000 miles per second. As the in-falling material smashes into the core, it causes the neutron matter there to rebound violently, sending a shock wave hurtling back out through the star. As the shock moves outward, it encounters increasing resistance and begins to slow down. Without help, it would fizzle out. But it is immediately followed by a flood of neutrinos, produced in the neutron core when it was squeezed. Neutrinos are massless, wraithlike particles that travel at the speed of light. Created during many types of nuclear reactions, they can pass right through an ordinary object like your body, or even the entire Earth, with little risk of being stopped. However, the matter in the decelerating shock wave of a giant star is so dense that it absorbs a significant number of neutrinos. This injection of energy into the shock wave gives it the boost it needs to finish the job of blowing away the outer layers of the star, rich in heavy elements.

The important point is that the properties of neutrinos are critically dependent on the strength of the weak force. If the weak force were just a little more feeble, then neutrinos would hardly interact with matter at all. In this case, the neutrinos from the core would virtually all escape without getting involved in pushing the outer layers into space. Conversely, if the weak force were a bit stronger, the neutrinos would be absorbed before they even left the core, rendering them impotent. Either way, the heavy elements manufactured inside giant stars would not become available for making into future planets, biochemicals, and lifeforms such as ourselves.

The catalog of cosmic coincidences has many other entries. For example, the manufacture of carbon inside stars calls for an especially fine piece of cosmic tuning. In this process, two helium nuclei have to come together to make a nucleus of beryllium, which then has to capture a further helium nucleus to complete the synthesis to carbon. But when the British astrophysicist Fred Hoyle first looked closely at this reaction in the 1950s he realized there was a problem. According to what was then known, the capture of a helium nucleus by a beryllium nucleus was far too unlikely to account for the observed cosmic abundance of carbon. He reasoned that the only way enough carbon could be made was if there existed a very specific match of nuclear energy levels—a "resonance"—between helium, beryllium, and carbon under precisely the conditions thought to prevail in the cores of stars at this stage in their evolution. Experiments promptly confirmed Hoyle's deduction; there was indeed a previously unsuspected resonance, very close to the energy value he gave. Crucially—for us—there is not a similar resonance at the same energy between carbon, helium, and oxygen. If there were, a large part of the carbon inside stars would quickly be changed into oxygen, and life as we know it would be impossible.

Once again, it seems, the universe treads a fine line. A deviation of just one or two percent from the actual value of the helium-beryllium-carbon resonance or the helium-carbon-oxygen resonance would have drastically reduced the amount of carbon around. As Hoyle himself remarked with characteristic Yorkshire bluntness: "If you wanted to produce carbon and oxygen in roughly equal quantities by stellar nucleosynthesis, these are the two levels you would have to fix, and your fixing would have to be just about where these levels are actually found to be. . . . A commonsense interpretation of the facts suggests that a superintellect has monkeyed with physics . . . and that there are no blind forces worth speaking about in nature."

Hoyle claimed that this "superintellect" had been at work elsewhere, tinkering with the emission frequencies of light from Sunlike stars, for instance, so that they match the absorption frequencies for photosynthesis in green plants. The properties of water—that substance so basic to all terrestrial life—also appear to be anomalous and (in Hoyle's words) "a put-up job."

Whether we choose to go along with such speculations or not, it is hard to escape the fact that there is something very peculiar and suspicious about all the many cosmic coincidences. We may stubbornly insist, of course, that without them we should not be here to marvel at how astonishingly well-tuned nature seems to be—and that is certainly true. But if the universe "just happened," it does appear extraordinary that it just happened to satisfy all of the exacting conditions needed to make life possible.

A number of leading scientists have given voice to this near mystical feeling that a life-enabling factor lies within the machinery and design of the world. The physicist Freeman Dyson, for instance, has commented: "It almost seems as if the Universe must in some sense have known that we were coming." And John Wheeler, too, has made the point: "It is not only that man is adapted to the universe. The universe is adapted to man."

This somewhat diffuse idea that nature is conditioned and constrained by our existence was first dealt with scientifically in a paper by the cosmologist Brandon Carter in 1974. He called the idea the Anthropic Principle and identified two specific forms of it. The least contentious of these, which Carter referred to as the Weak Anthropic Principle (WAP), simply affirms that those properties of the universe we can make out (including the observed values of all physical and cosmological quantities) are self-selected by the fact that they have to be consistent with our own evolution and present existence. All the WAP really does is draw our attention to the restrictions placed on properties such as the size, age, and laws of the universe by our presence—a statement that for many astronomers and physicists is quite anthropic enough. Indeed, the majority of physical scientists regard any hint of intrusion by man and mind into cosmological affairs as anathema. We are, after all, so small and new while the universe is so very large and ancient. What possible significance could we have on a cosmic scale? Yet, against all the odds, nature meets every one of the stringent conditions needed to create carbon-based life and consciousness.

It was to address the extreme unlikelihood of these coincidences that Carter put forward the alternative Strong Anthropic Principle (SAP). According to this, the universe *must* have those properties that allow life to develop within it at some stage in its history. Whereas the WAP only asserts that the constants and laws of nature are such as to admit life (a fact which is beyond dispute), the SAP goes much further and insists that they could not have been any other way. In effect, the SAP maintains that only a universe with intelligent observers has any meaning and, furthermore, that observers can only arise given a unique set of cosmic parameters and physical laws.

Since its first enunciation by Carter, the Strong Anthropic Principle has been widely debated and variously condemned, condoned, and elaborated. In particular, it has been used by some researchers as a springboard for further and more exotic speculations. Of these, none is so intriguing or far-reaching as John Wheeler's Participatory Anthropic Principle (PAP). The thrust of the PAP is that we—together with any other conscious observers who may exist in space and time—are the necessary and sufficient means by which the universe is brought into being. A possible mechanism for such observer-created reality has already been seen: the collapse of the wave function during observation in quantum mechanics.

According to the PAP, each conscious observer contributes a little to making the universe real through the collapse of wave functions. But if this is true, then there must inevitably be observers in the future helping to make *us* real since, following Von Neumann's analysis, a valid observing system cannot bring about its own collapse to a particular quantum state. Carrying this argument indefinitely forward in time (as the *Scientific American* columnist Martin Gardner first pointed out) leads to the Final Anthropic Principle (FAP). This states that not only must intelligence evolve in the universe, but once it has come into existence it will never die out. It must continue to live and grow without limit so that it can participate in the creation of everything that came before it.

Wheeler draws a useful analogy between his participatory cosmos and a variation of the old parlor game Twenty Questions. At a party in which guests were playing this game, it was Wheeler's turn to leave the room while the others agreed upon a new word. Upon Wheeler's return, each guest in turn replied to his questions with the customary "yes" or "no." To his puzzlement, however, Wheeler noticed that the gap between question and answer seemed to take longer and longer as the game went on, as if the respondents were having to do an increasing amount of hard thinking. Eventually, he guessed "sky," to which, after an inordinately long pause, came the reply "yes!"—accompanied by a burst of laughter. Wheeler was then let in on the secret: there had been no word chosen to begin with. Instead, each reply to his question was given arbitrarily and used to narrow down the range of possibilities until, in the end, there was only one subject consistent with all the previous answers—sky.

Might it not be the same with the universe, Wheeler asks. Might it not be that mind effectively "asks questions" of nature by observing it, while to each observation the universe "replies" by manifesting itself in one particular way—the only way that is consistent with the eventual evolution of a sentient mind? If so, then as time goes on, reality becomes ever more sharply defined until there is just a single possible final "answer"—a unique universe, past, present, and future whose internal logic has to be such as to allow creation by intelligent observership.

Clearly, the PAP posits an extravagantly elevated status for mind. It forces us—the human race, our offspring, and any other self-aware beings there may be in space—onto center stage, where we are accustomed to be. So, not surprisingly, this is a very controversial theory. A typical response is that of the leading theoretical physicist Roger Penrose: "The circularity and paradox involved in this picture has an appeal for some, but for myself I find it distinctly worrisome—and, indeed, barely credible." We should, quite rightly, be wary of uninhibited anthropocentrism.

But what are the alternatives? They are that the universe was created by a god—an unknowable, supernatural being—or that the universe came about by blind chance. Both of these ideas, though at first sight apparently more "reasonable" (and certainly more conventional) than Wheeler's participatory theory, suffer from troubling difficulties of their own.

Beliefs in a deistic creation, of course, are very old and take many forms. At their common core, though, is the assumption that an intelligent being (or collection of beings) was the fundamental cause of the cosmos in which we live. In some theologies this being continues to play an active and interested part in running the universe on a day-to-day basis, whereas in others He merely supplies the genesis spark and then stands back to let nature take its course. Most of us grow up inculcated to a greater or lesser extent with this sort of viewpoint, so that, even if we disagree with it, it seems fairly natural and commonplace. But it is anything but that. The notion that some ancient, omnipotent creature designed and fashioned the whole universe from scratch is actually very extraordinary indeed. Considered dispassionately, it is really far more bizarre than Wheeler's pantheistic concept of a universe that is self-made, because at least Wheeler's scheme offers a visible (albeit controversial) creative agent and mechanism—namely, ourselves. Religion, by contrast, expects us to have faith in something we have never seen and, we are told, can never hope to understand.

Are we just to accept blindly that God made the universe, without inquiring who or what may have made Him? If so, then we have answered nothing. God is just a question mark—simply another way of saying we don't know where the universe came from. As the distinguished science essayist and geneticist J. B. S. Haldane once wrote: "The theory of [a deistic] creation is essentially a refusal to think back beyond a certain time in the past when it becomes difficult to follow the chain of causation. To hold such a belief is, therefore, always an excuse for intellectual laziness, and generally a sign of it."

But if religious cosmologies have their faults, then so

too do scientific ones. In fact, many of the latter are as evasive as their theological counterparts when it comes to the question of ultimate origins. Standard Big Bang physics has had remarkable success in describing the likely contents and environment of the infant universe back as far as the first trillion trillion trillionth of a second. But it runs into major difficulties in its attempt to extend our knowledge all the way back to the very beginning—to "Time Zero" itself. There is a huge gulf—a philosophical as well as a physical gulf—between identifying subsequent causes and effects and uncovering the nature of the very first event of all.

Fashionable with today's cosmologists is the idea that the universe arose initially ex nihilo—out of nothing. In a classical, commonsense world, that would be an absurdity (barring supernatural intervention); there would be no possibility of nothing ever becoming something. But quantum mechanics flies in the face of reason and insists that unpredictable fluctuations in physical systems can indeed take place—that there can be causes without effects. The extent of these random fluctuations is under the jurisdiction of the so-called uncertainty principle, discovered by Werner Heisenberg. This puts strict limits on how accurately we can pin down the values of certain pairs of quantities. For example, we can never know precisely both the momentum of an electron and its position at the same instant. The more closely we try to fix one quantity, the more the Heisenberg uncertainty principle throws the other quantity out of focus. A similar conjugate relationship exists betwen energy and time.

One way to interpret the energy-time uncertainty is to imagine a tiny volume of "empty" space. In Newtonian physics, such a region would be absolutely and permanently devoid of energy or matter. But quantum mechanics insists that nothing is assured. Even in a perfect vacuum, a certain amount of energy, E, might come into existence, provided it does so for less than a certain time, t. The uncertainty principle defines the relationship between E and t; the bigger E is, the smaller t must be. That is to say, energy can spontaneously appear from nowhere as long as it disappears again quickly, before any measurement can detect it.

Einstein showed that energy and matter are equivalent and interchangeable: matter is like frozen energy. So the uncertainty principle effectively allows particles of matter to appear out of empty space, provided they smartly vanish again before the rest of the universe notices them. Such unseeable, transient, or "virtual" particles are necessary, it turns out, to explain the finer details of how the electromagnetic force works. In this sense, they have been shown to be "real."

Now, if virtual particles can appear out of a perfect vacuum today, it may be, the argument goes, that they also appeared out of the nothingness that is supposed to have existed before there was a universe. This is the basic tenet of creation ex nihilo in its modern, quantum-mechanical form. In the beginning, there was nothing—no matter, energy, space, or time. Then, because of the "fuzziness" allowed by the uncertainty principle, there was a random fluctuation. A bit of matter spontaneously appeared. Normally, this would have disappeared again almost immediately. But in the instant after it came into existence, other processes occurred (to do with the splitting of the four basic forces from a single original "superforce") which triggered a brief but exponential burst of expansion. During this so-called "inflationary" period (lasting about a billion trillion trillionth of a second), the infant universe ballooned from well under the size of a proton to about the size of a basketball. The physics and physical conditions involved were extremely esoteric. But the net result was that at the end of the inflationary epoch a kind of change of state took place, similar to when water freezes, in which a vast amount of energy was suddenly released. This energy went into making all of the subatomic particles the universe would ever contain.

At first sight, it seems a cogent and wonderfully economical idea. The universe, as one researcher put it, "may be the ultimate free lunch." But a closer examination of the ex nihilo theory reveals some serious, and perhaps fatal, flaws. First, there is the ontological point that a timeless, spaceless "nothing" cannot ever really be said to exist, because it is the very negation of existence. At best, nothing can exist for zero time—a situation which, even the most ardent nihilist must admit, is the practical equivalent of nonexistence. Since nothing cannot exist, any theory that employs it as the basis for generating something effectively undermines itself from the start.

Second, there is the problem of the origin of time. The idea that time and the universe may have originated together is not new; it was first suggested by St. Augustine in the fifth century, and has now merely been revived in modern Big Bang cosmology. But it is difficult to see, according to any version of this theory, how the universe could possibly have got off the ground if it was in a timeless state to begin with. Any form of activity—even quantum mechanical—must take place *in* time. This crucial point seems to be consistently (and conveniently) overlooked by cosmologists when they refer to the supposed creation of space-time. Time cannot be created, because creation itself is an activity that takes time in which to happen. The only reason the uncertainty principle—along with the rest of physics—can operate today is that it has a preestablished temporal setting. In a genuine state of timelessness, there can, by definition, be no change, so that it would be impossible to make the transition from some precosmic condition to the infant universe.

Third—and this is a related point—if the uncertainty principle is invoked as the instigating mechanism for creation, then it must be assumed to have already been in existence at (or "before") Time Zero. But then where did the uncertainty principle come from? We have merely substituted one origin problem (that of the physical universe) for another (that of the laws of nature). It is far from clear how, in a supposedly immaterial, spaceless, timeless state, the un-universe could have selected and acquired one particular set of rules to govern its future activity over all others.

Each one of these criticisms is by itself enough to threaten the theory of creation ex nihilo. But together they are very damaging indeed. Perhaps it is time for those who favor the idea of a unique genesis event—and, in particular, that space and time were created at some specific moment in the remote past—to look more closely at the serious logical and philosophical inconsistencies of such a belief.

We need to consider, too, the unsatisfactory way in which mainstream cosmologies—both religious and scientific—address the issue of cosmic coincidences. Theologists might say that God fine-tuned the universe to make it suitable for human life (in which case, incidentally, they ought not to balk at the Participatory Anthropic Principle for being overly anthropocentric). But this simply raises, again, the problem with all Designer arguments: their failure to account for the nature and prior existence of the Designer. Under what set of physical laws did God operate before he made the universe—and where did those laws come from? If He was intelligent enough to engage in large-scale cosmic engineering, by what means did He evolve that intelligence? How did He know that the system of natural laws He chose before the Big Bang would inevitably lead to thinking creatures like ourselves? How could He simulate the cosmos before there was a cosmos? And if He existed before time, then how did He act and plan and make ready for His genesis experiment?

The scientific ex nihilo proponents, on the other hand, can "explain" cosmic coincidences—but only by resorting to a cosmology based on Everett's Many Worlds Interpretation. That is, they are forced to assume that there was not just one but an infinity of universes created in the beginning by quantum fluctuations. We just happen to live in the one that is conducive to human life—so naturally we think it is special. But there are several grave problems with this idea. First, it continues to rely on the prior existence of the uncertainty principle as a creative influence for bringing space and time into existence—an idea that seems fundamentally flawed. Second, it substitutes the supposed implausibility of observer-made reality for the even greater improbability of an infinitude of universes. (The introduction of infinity is also a major handicap to theories that invoke endless cycles or oscillations of the universe to avoid a genesis moment.) And finally it asks us to believe in the existence of other space-times of which we have no knowledge nor any possibility of ever acquiring it.

The fact is that no cosmological theory currently on offer is free from detractors. The mainstream cosmologists would like to convince us, and themselves, that they are close to revealing the very mechanism of genesis. But the truth is that they are merely refining their understanding of the causes and effects that came after the instant of the Big Bang. They are getting no nearer at all to isolating the ultimate mechanism of creation.

This is precisely why we should begin to treat very much more seriously the possibility of a link between mind and cosmology. Through the Participatory and Final Anthropic Principles we do have a way to account for the creation and existence of the universe without resorting to a unique genesis event. Creation and the selection of a specific, intelligence-generating universe can go on continuously, so avoiding the need for a logically inconsistent timeless state. Furthermore, these extensions of the Strong Anthropic Principle provide the opportunity of a remarkably comprehensive picture, including answers to such profound questions as: Why are we here? What is the "purpose" of life and consciousness? And what is our ultimate destiny?

The main criticism of the PAP and the FAP have been that they call for apparently unreasonable powers of the human mind. But just because we seem unimportant today does not mean we shall always remain that way. After all, we only began our descent from the apes some five million years ago. We have only been truly human for perhaps the last third of a million years. And consider how incredibly recent is our culture and technology, even compared with these surprisingly brief biological time scales.

All eternity stretches before us, so if we are already playing a part—however small—in deciding what is real, then how much greater will our influence be in a million or a billion years? We are only a tiny part of what the human race may become in the millennia and eons ahead. Whatever lies in store, whatever the final answers to the universe's deepest mysteries may turn out to be, the journey of mind has only just begun.

Part III

MIND

Not once in the dim past but
continuously through the conscious
mind is the miracle of
creation wrought.

—SIR ARTHUR EDDINGTON

10

To Distant Shores

Just as erect man left his African homeland a million years ago, so his descendants are now poised to move beyond their home planet. We have spied the landmarks of space through our telescopes, dispatched robot scouts to map out the neighboring regions of the solar system. And now it seems it will not be long before we follow in person. A new, dramatic phase in the spread of consciousness and intelligence is about to begin.

How this latest odyssey of the human race might unfold we can only vaguely conjecture. As one of the most successful scientific prognosticators, Arthur C. Clarke, has pointed out, "It is impossible to predict the future, and all attempts to do so in any detail appear ludicrous within a very few years." The most we can reasonably hope to do is define the limits within which the possible futures of man and mind may lie. What are the barriers we face in our efforts to colonize space? And what technologies and physical processes might we exploit in bridging the unimaginably large gaps between ourselves and the stars?

The present signs are that over the next few decades we may actually see a reduction, or at least a leveling-off, of manned activity in space. This will happen as we shift our attentions instead to more urgent ecological and social issues. During the resulting temporary lull we shall have the opportunity to step back, review our successes and failures to date, and formulate a more coherent space policy for the future that goes beyond mere national interests.

The first goal of any revitalized manned program (as the world's space agencies have long recognized) must be to secure a permanent foothold in orbit. From small, research-oriented space stations we must try to move smoothly and progressively to larger multipurpose platforms. Only close international cooperation, both on the Earth and beyond, will make such projects possible. At the same time, in becoming familiar with the sight of our small world from afar, we may come to place more emphasis on the commonality of man rather than our political or ethnic differences.

With the easing of tensions between East and West, it may be that some of the remarkable devices spawned under heavy secrecy within our major defense laboratories will become available for peaceful uses in space. Already, for instance, suggestions have been made for adapting various components of the Strategic Defense Initiative (SDI) for future interplanetary and even interstellar exploration.

The first of these, put forward by engineers at the Jet Propulsion Laboratory (JPL) in Pasadena, California, concerns an advanced weapons system, still in its early stages of development, known as the rail gun. Basically this works like a particle accelerator, by feeding brief but intense electrical pulses to a long chain of electromagnets and thereby accelerating objects to high speed. But in contrast to a conventional accelerator, which produces a subatomic beam, the rail gun is designed to fire out macroscopic projectiles. In their "Star Wars" role, these projectiles would be shot from orbit to intercept and destroy enemy missiles in flight. But the JPL scheme proposes a very different use for the rail gun—as a means to launch tiny spacecraft.

A two-pound probe, about the size of a coffee jar, could be hurled so quickly from a rail gun that it would reach Saturn in just twenty-four months. That is less than half the time it took the highly successful Voyager craft. The rail gun also has another potential advantage over present-day launch methods: its military specification calls for it to be able to fire at least twice a second. If miniature probes could be mass-produced, there is nothing to prevent fifty of them, or even a thousand of them, from being dispatched in a single year. Nor would their very restricted mass—about nine hundred times less than that of the Voyager probes—necessarily pose a problem. If our present rate of progress in miniaturization is maintained, we can look forward to some remarkable technical advances by the middle of the next century. Much of the equipment needed for spacecraft navigation, data collection and transmission, and power generation will be dramatically reduced in size and mass, as a result of which very small (and inexpensive) spacecraft—microprobes—will become a real possibility.

One obvious function of microprobes would be to serve as the van for more sophisticated spacecraft by mapping and carrying out preliminary examinations of new worlds and environments. They would enable not just every planet and moon, but every major comet and asteroid as well, to be surveyed, sampled, and tested within a relatively short period. Further technical advances might allow microprobes to become fully autonomous laboratories, equipped for decision making and able to be launched at much higher speeds. An initial speed of, say, fifty miles per second would put Jupiter a mere four months' journey time away—and even Pluto could be reached in five years.

Just one problem might remain unsolved: how to slow such a probe at its destination. For all its sophistication, it would be surprising if twenty-first-century technology managed to find a way to build minute yet powerful rockets that consumed virtually no fuel. The tiny onboard thrusters of a microprobe would probably only suffice to nudge the spacecraft into its final orbit. For the bulk of its deceleration, a microprobe would still rely on the gravitational currents between worlds—specifically, on the boundary layer that exists between the opposing pulls of two gravitating objects. If the initial course and speed were precisely set, then the little craft could bob and drift in the interplanetary void to be washed up within meters and milliseconds of its intended goal. Of course, probes might go astray while the technique was perfected. But steadily the success rate would improve, until the deep-space tracking stations of

Earth were inundated daily with data from thousands of these small explorers all over the solar system. Microprobes, or other robot craft propelled perhaps by solar sails or nuclear engines, will be our envoys and scouts: the trails they blaze, for us to follow.

Man first penetrated space, it is true, in the twentieth century. Yet there are too many urgent problems culminating at present for us to build securely on our early successes. Not the twentieth, but the twenty-second century may prove to be the true dawn of the Space Age, for it may only be by then that we are able to put the infrastructure for future colonization in place. It is simply not enough to have one or two small orbiting space stations, an occasional satellite launch, a tiny fleet of manned space shuttles, and a commitment that ebbs and flows with the political tides. Nor is it enough to achieve some high-profile objective, like landing astronauts on the Moon or on Mars, without this being part of some more far-reaching program. The momentum, the presence, the whole technological foundation of space exploration has to be sustained and extended indefinitely. Success has to build upon success, just as in climbing skyward a rocket must fire continuously, gathering speed, in order that gravity does not drag it back down.

What man must strive for is a sort of critical mass of space involvement to ensure that the process never stops. Then, even a disaster, such as the total loss of a crew or of some elaborate piece of hardware, could be absorbed. And while there would doubtless be an inquiry into such a tragedy, it would no longer be enough to halt or even seriously slow the rest of the program.

When there are substantial colonies in orbit and on the Moon and on Mars, not all of the impetus for further exploration will necessarily derive from Earth. Eventually, Mars may have its own plants for manufacturing spacecraft, its own interplanetary launch pads. From the Moon, too, missions may depart for more distant worlds. Later, there may be mining operations in the asteroid belt, supplying fuel and building materials to nascent communities around Jupiter. Mars may become self-governing and self-reliant, no longer a dependency but a partner of Earth. And so, as travel between the worlds of the solar system becomes increasingly routine, as families emigrate from Earth to the Moon, or to distant Ganymede or Europa, mankind will contemplate its next great move.

It is not hard to reach the stars. We have already proved that. Four probes launched in the 1960s and '70s (Voyagers 1 and 2 and Pioneers 10 and 11) are well on their way to breaking ties with the Sun and drifting free in interstellar space. Merely reaching the stars is no problem at all. It is reaching them in a reasonable length of time that poses the massive challenge. The technological gulf between the Wright Brothers' seminal flight and Voyager 2's spectacular flyby of Neptune may seem great. But it is dwarfed by that which separates Voyager from the first practical starship. To send even a small robot probe to the nearest star within a human lifetime calls for the orchestration of forces that not even twenty-second-century man may fully tame.

And there is another factor. What perverse logic would it be to dispatch a star probe now, or in the near future, knowing that within a few decades it might be overhauled by a craft of much swifter design? Given that new forms of propulsion are inevitable, it would be foolish to strain to reach the stars by less-than-adequate means, especially when there is still so much to be achieved in the solar system.

If mankind learns anything over the next two hundred years, it will be to move patiently, rationally, and with some grace toward its highest goals. So, possibly three or four more human lifetimes will pass before the first purpose-built star probe slips effortlessly from the Sun's grasp, bound for one of our near stellar neighbors.

And in this venture, it may be that a second offshoot of today's Strategic Defense Initiative finds an eventual, peaceful use: the high-energy laser. This is a beam that could instantly vaporize ICBMs or even entire cities, but that in time might shine for a very different purpose—across the lonely light-years to push a spacecraft to the stars.

Imagine: a gleaming disk, ten miles across, made of aluminum half a millionth of an inch thick. It is a giant sail to catch a wind of light. And though its mass is close to a ton, the overwhelming bulk of this is focused at the center in a tiny riot of complexity—the instrument cluster. Such may be man's first true star probe, and this the moment of its departure.

Somewhere between the orbits of Saturn and Uranus lies the means to propel the strange craft on its long voyage. It is a 1,000-trillion-watt laser, powered by fusion reactors. Stationed in front of this is a lens 700 miles wide to focus the brilliant light.

In a soundless concussion the laser erupts into life. Its beam is trained unerringly upon the probe's huge sail, impelling the craft forward, driving it on with an acceleration 150 times that of Earth's gravity, so that within less than three weeks the starship's speed is half that of light.

And then, just as abruptly, the laser shuts off. And the craft begins its coast to . . . where? Maybe to Tau Ceti: a nearby orange dwarf that, some suspect, harbors its own sundry collection of worlds.

An ingenious method is used to slow the probe as it nears its destination. The slender metallic sheet that mirrors the distant laser's light is made, it turns out, in two segments. There is an outer "decelerator" stage, the full ten-mile width, and an inner "rendezvous" stage spanning a third of that diameter. As the probe closes to within half a light-year of Tau Ceti, the outer sail cuts loose from the inner, while the inner swings around so that its reflective surface now points away from the Sun. Twelve years earlier—at the halfway point in the voyage—the far-off laser was reactivated so that its light, now arriving at and striking the outer sail, would rebound onto the pursuing inner stage. It is as if a light beam from Tau Ceti itself were acting to brake the craft's rendezvous stage. And in this way the core of the craft, bearing all of the instruments needed to explore the new worlds of this unfamiliar star, is gradually slowed. Finally, the central payload can break away altogether, maneuvering itself with small conventional thrusters toward its final goal.

By such means may man, vicariously at first, reach the stars. In a sense, it will be a homecoming since it was of ancient star flotsam that the Earth was made—the Earth and the living things that by slow stages have evolved and grown upon it. Now that long-lost stellar matter has intricately rearranged itself. Over five billion years it has come together to make oaks and apes and, eventually, humans whose brains can question their own origins, purpose, and destiny.

Even as the warming rays of an alien sun fall upon our first interstellar probe, similar craft may be setting out for other nearby stars: to Alpha Centauri, triple-sunned neighbor of Sol; to dull-red 61 Cygni; to brilliant Sirius with its white-dwarf mate; and further afield to Sigma Pavonis, so like the Sun. If we follow the pattern of our present-day planetary expeditions, we shall go on to explore, map, and survey much of our interstellar neighborhood automatically with self-controlled machines before ever attempting missions that involve human crews.

Sending astronauts to the stars, in fact, poses enormous technical difficulties. A manned starship must carry ten thousand times or more the payload of a mere robot probe. And a human crew, too, should at least have the option to return. What kind of laser could blow a ship so large to even the nearest star and back in good time? It may be possible in theory, but we need to look very carefully at alternatives that might be more practical.

The laser sail was conceived in the first place to sidestep a monumental problem in rocketry: if the propulsion system and propellant are carried on board a spacecraft, then the total mass will be greatly compounded. More mass demands yet more propellant to achieve the same acceleration, which increases the mass further, and so it continues in an upward spiral. But worse, to lose speed is as hard as to acquire it. And—the crux of the problem—carrying the necessary propellant for deceleration does not merely double the difficulty of the mission, it squares it.

For instance, it might take a million tons of fuel to reach an acceptable cruising speed. But to cancel that speed at journey's end, the craft would have to set out not with a million, or even two million, but with an outrageous million million tons of propellant. And if the crew wished to return, we should have to square this value again.

That is why the rocket engines that first bore humans to the Moon could never take them on to the stars. Chemicals are far too ponderous for the amount of thrust they produce. Even the more advanced nuclear and ion motors, potentially so useful for interplanetary flight, could never furnish a stardrive.

But there are other possibilities—such as antimatter. Mixed even in tiny amounts with ordinary matter, it annihilates completely, liberating huge amounts of energy. In fact, it releases ten billion times more energy than that which comes from burning the same mass of high-grade chemical fuel. Just a few precious tons of antimatter would suffice to push a manned starship to about one-tenth the speed of light, after which a different type of propulsion system would be needed to achieve further acceleration. One possibility for this second-stage "hyper-drive" is the so-called interstellar ramjet.

It is a dream as old as space travel itself: to harvest the ultra-thin gas between the stars as a source of fuel. Gathered up in sufficient quantity, those sparse interstellar particles—mostly of hydrogen and helium—could power a starship to almost any speed, even to within a whisker of that of light itself. An interstellar ramjet would suck in charged particles, or ions, of hydrogen and helium with a magnetic field produced by a colossal funnel-shaped scoop attached to the front of the spacecraft. (It could only do this efficiently when traveling at a significant fraction of the speed of light—hence the need for another kind of motor, such as an antimatter rocket, to provide the initial boost). The harvested ions would then be fed as fuel to a nuclear fusion reactor. Such a reactor would create a high flux of very fast-moving particles, which, when allowed to escape from the rear of the vehicle, would provide the forward thrust.

Sketched out in this way, the interstellar ramjet may seem like a very practical and feasible way to achieve high-speed interstellar travel. Yet, in fact, it poses a number of formidable technical problems. No one yet knows, for instance, how a magnetic field could be generated ahead of the ship large enough to pull in the necessary amounts of fuel. Even more perplexing is how we could extract sufficient reactive thrust from the harvested gas in a fusion reactor. The answers may lie far beyond the ken of present-day engineers, but they are less likely to elude our descendants several centuries from now.

It is hard to imagine us turning our back on the challenge of the stars. We may be delayed somewhat by our current terrestrial crises. We may take some time to solve the technical difficulties. But we are too inquisitive, too driven by a relentless urge to explore, to remain forever closeted in the solar system. By some means, if we do not destroy ourselves first, we shall almost certainly achieve interstellar flight. And in starships that travel close to the speed of light, we should eventually be able to travel to other suns well within the confines of a human lifetime. The Einsteinian effects, which alter the relationship of any fast-moving object to its surroundings, make that a possibility. Time aboard a spacecraft moving at 99 percent of light-speed, for example, would run seven times more slowly than it does on Earth. So that during a journey of seventy light-years, travelers might age by only a decade while their relatives at home went almost from cradle to grave.

Bizarre as it may seem to us now, such time distortions are likely to become an accepted feature of future interstellar travel. Seasoned voyagers between the stars may watch their planet-bound children, and even their grandchildren, gradually surpass them in chronological age. That may be an unthinkable situation today, but attitudes will inevitably change over the next few centuries—and not just attitudes. For the fact is that human technology will no longer be confined to toolmaking: to the building of devices, however elaborate, outside of man.

The gradual conquest of all major forms of disease is already under way. At the same time, advances in surgery promise the eventual easy and safe replacement of any worn-out body part. The process of aging may be halted or greatly slowed through genetic engineering (though one negative consequence of this would be to greatly exacerbate the population problem). People may someday live for two or three hundred years, and perhaps more, so that, in view of the ever-increasing speed of spacecraft and the subsequent contraction of shipboard time, journeys to other stars will consume a smaller and smaller fraction of the average human lifetime. As a result, such journeys will become increasingly routine, moving people in and out of the standard, stationary time stream, aging them at different rates, scrambling the generations.

Looking at prospects for the more remote future, we can envision a time when man's domain spans a thousand light-years or more in every direction from the Sun—a vast bubble of space, home to tens of thousands of stars and their numerous, colonized worlds. We can hardly begin to imagine what startlingly varied cultures, philosophies, and art forms might spring up as our offspring adapt to all of these alien environments.

There is the possibility of meetings with other intelligences, though how we would react, especially to finding life-forms very much more advanced than ourselves, is hard to say. If we encounter creatures whose technology is a thousand or a million years ahead of ours, what effect might that have on us? Would we rapidly, hungrily assimilate their knowledge, leaping forward until we were their intellectual equals? Or would the realization that we had evolved so relatively little crush us and stifle our ambitions? A child is not put off by the fact that it knows less than its parents or teachers. On the other hand, we have become so accustomed, as a race, to having no peers—let alone superiors—that the discovery of high alien intelligence might come as a profound and disabling shock. The best we can hope for is that all beings capable enough to journey to the stars are also mature enough to be able to coexist in a spirit of peace and friendship. If this is the case, then it might represent a further and most important opportunity for the evolution of conscious thought.

11

Masters of Space and Time

The speed of light—186,282 miles per second—represents an inviolate upper limit to the rate at which we can ever hope to travel conventionally through space. This is because the mass of any material object—and therefore the effort required to make it go faster—tends to infinity as the speed of light is approached. Although, in principle, the relativistic slowing of time still allows very long journeys to be made in arbitrarily short periods of time, there is a serious drawback in that the rest of the universe ages a great deal faster than any near-light-speed travelers. For example, a crew might reach the center of our Galaxy within a few months aboard a fast enough spacecraft, but those who did not make the journey would have aged by thirty thousand years. Such massive dislocations in time would be hard to accommodate even within a highly advanced society. In fact, the more that human consciousness spreads out from the Earth the more it will be in danger of being inhibited and fragmented by the finite speed at which anything or any information can travel.

There is possibly a way, however, in which this cosmic speed limit could be circumvented. A time may come, many centuries from now, when man no longer has to move or communicate exclusively through normal space. The alternative methods he may turn to are clearly a matter of extreme speculation today. But their consequences—particularly for the diffusion of human awareness and intelligence to remote parts of the universe—could be so extraordinary that we cannot easily dismiss them. We must look at the very real possibility that, in the far future, our progeny may learn how to manipulate space and time.

The things we call space and time, Einstein showed, are inextricably woven together to form a flexible, deformable space-time. Space-time can be pictured in some ways to be like a rubber sheet. In places where there are no objects—no material things—the sheet is perfectly flat. But where a mass sits on the surface of space-time it causes a dip—the greater the mass, the greater the dip. Our imagination is helped, of course, by the fact that the surface of a rubber sheet has only two spatial dimensions and bends through a third. "Flat" space-time, on the other hand, consists of three spatial dimensions (plus one of time) and so must bend through a fourth. This fourth, and—to us—

inconceivable spatial dimension, is known as hyperspace.

A mass that is highly compressed (imagine a cannonball on a trampoline) creates a hollow in space-time that is very deep and steep-sided. In fact, if the mass is sufficiently concentrated, its space-time hollow will have such steep sides that not even light can climb out of it. Such escape-proof regions of space-time have been dubbed, for obvious reasons, black holes. They are thought to exist in at least two places: the sites of very massive stars that have exploded as supernovas and the centers of large galaxies. From our point of view, they are important because they might just furnish a remarkably effective means of getting around the universe.

In theory, black holes can possess only three intrinsic properties: mass, spin, and electric charge. Also in theory, what we might call the "internal space-time architecture" of a black hole depends crucially upon which of these properties are actually present. A black hole that only has mass is effectively a dead end; nothing that falls in can avoid being crushed out of existence upon arrival at the center—the so-called singularity—where the gravitational force is infinitely high. But it is difficult to see how such a simple black hole could ever come about in the real universe. All objects in space, including planets, stars, and galaxies, rotate, so that a black hole that formed from all or part of such an object should also rotate. (We can ignore the possibility of charged black holes since rotation is likely to be a far more important factor.)

A spinning black hole has a much more complex and interesting internal architecture than the plain, mass-only variety. In particular, theory suggests there are possible pathways through a spinning black hole that avoid any dangerous singularities and that travel instead along a space-time tunnel known as a "wormhole." The wormhole, in turn, leads back out into the normal universe again—but not into the same part of the universe as the opening to the black hole. The far end of the wormhole, astonishingly, may lie millions or billions of light-years away and be displaced by huge periods of time from the entrance.

At least on paper, a wormhole provides a shortcut between two otherwise widely separated regions of space-time. It would be as if the universe were folded over in such a place. A journey between two remote points through conventional space-time—that is, around the curved surface of the fold—would be as long as if the space-time between the points were flattened out. But a journey through a wormhole connecting the two points directly, outside of normal space and time, could be achieved almost instantaneously.

None of these ideas, surprisingly, are new. The popular terminology may be modern, but scientists have known about the possibility of wormholes for more than seventy years—since shortly after Einstein put forward his general theory of relativity, which related gravity to the geometry of space and time. A wormhole, mathematically speaking, is a solution to the field equations of general relativity, in which two regions of the universe are connected by a short, narrow "throat" (known formally as an Einstein-Rosen bridge). The problem, from a practical standpoint, is that such a structure is inherently unstable. Any matter or radiation that falls in is so concentrated and amplified by the gravitational fields of the wormhole that its own gravity alters the internal space-time architecture and immediately closes off the hole. Moreover, a wormhole exerts gravitational forces as strong as the black hole itself, with the unfortunate result that a wormhole of moderate size—say, from a few meters to a few kilometers across—would tear any travelers of human dimensions to shreds before they even got near it. Clearly, some artificial modifications would be needed before wormholes could be opened to the public as a sort of cosmic subway.

Science-fiction writers—ever heedless of technical trivia—have been whisking their heroes effortlessly around the universe with the help of wormholes for a number of years. Carl Sagan, for instance, though not the first to do so, exploits this method in his novel *Contact*. In fact, it was Sagan's well-publicized yarn that prompted Michael Morris and Ulvi Yurtsever, together with their Ph.D. thesis adviser, Kip Thorne, at the California Institute of Technology, to begin looking for new features of wormholes in 1985. They wanted to investigate ways in which wormholes might be used for practical travel through space and time.

What the Caltech team did was to build mathematical wormhole geometries in which it was specified that the throat stayed open and the gravitational fields were modest enough to allow a safe human passage. Given these constraints, the equations of general relativity then prescribed what kinds of matter and energy were needed to generate the holes. The researchers found that the throats of their holes had to be threaded by matter or gravitational fields with enormous negative pressure. That is, the matter sustaining the hole would have to be under tension, like a stretched spring. For a hole about a kilometer across, the tension is similar in magnitude to the pressure at the center of a neutron star (a star slightly more massive than the Sun but with a volume roughly equal to that of a large mountain). For a smaller hole, the tension would be even greater. Most crucially, in all cases where the wormhole remained passable, the tension would be greater than the energy per unit volume, or energy density, of the matter itself—a very peculiar condition.

In ordinary matter, tensions and pressures are always much less than the energy density. For example, the breaking tension of steel is about a trillion times smaller than the energy stored per unit volume of the metal. Matter in which the internal tension exceeds the energy density is said to be "exotic." It is this exotic matter, according to Einstein's field equations, that is needed to make a traversable wormhole.

Does exotic matter exist anywhere in the universe—and if not, will we ever be able to manufacture it? The simple answer is that no one knows. However, it would be surprising, given our increasing ability to make all sorts of new and unusual subatomic particles in laboratories today, if we did not find some way to bring exotic matter into being. It would then be a question of developing suitable varieties that interacted weakly enough with other matter to avoid harming a traveler and that could be confined within a wormhole away from the path down which the traveler must go.

Spurred on by the Caltech team's encouraging results, other theorists have joined the investigation of traversable wormholes. Matt Visser of Washington University, St. Louis, for example, has found a kind of wormhole so benign that travelers could pass through it without encountering any matter, exotic or otherwise, and without feeling any forces at all. The extraordinary and potentially lethal forms of matter and energy needed to keep the wormhole open are completely isolated from the traveler's permitted flight path.

At least theoretically, then, the door seems ajar to the prospect of this fabulous means of transport. All we need to do is set up one end of a wormhole near to the Sun and arrange that the other end opens out somewhere close to our desired destination—perhaps another star inside our own galaxy, or perhaps another galaxy altogether. At the entrance to the wormhole would be a black hole; at the exit would be—logically enough—a "white hole" (the time reversal of a black hole). We would disappear down the black hole, travel along the wormhole and emerge, a short time later, out of the white hole at our faraway port of call. Since it would be impossible to return by the same route (you can only go in black holes and out white holes), we would have to establish a second wormhole parallel to the first but that permitted movement in the opposite direction through space and time.

Continuing this speculation, we can imagine a growing network of such wormhole pairs linking more and more star systems and galaxies. Eventually, large, "high-volume" wormholes (the equivalent of freeways) might link major objects such as galaxy clusters, while branch lines radiated out from the main terminals within a cluster to individual galaxies and then to individual stars.

The likelihood is that we would have to construct all of this network from scratch, since the chances of natural black holes and wormholes being in handy locations is not very great. To give an example, the nearest object to the Earth strongly suspected of harboring a black hole is the X-ray source Cygnus X-1, which lies about ten thousand light-years away. It would obviously be very inconvenient, even using high-speed starships, to have to travel so far through conventional space (cutting ourselves off from our own time) to catch the next wormhole express.

The evidence for black holes is compelling, but it also suggests that they are not sited where we could reasonably use them as a means of transport. Moreover, a natural black hole, and any wormhole associated with it, would almost certainly not be traversable, nor, if it were, would we have any way of knowing where it led to.

One of the most intriguing possibilities, suggested by Morris, Yurtsever, and Thorne, is that we could "grow" our own large-scale holes. One of the consequences of recent attempts to link the quantum mechanical view of the world with general relativity is that, on the smallest of scales, space-time can be treated as if it were a sort of continuously shifting foam of wormholes each measuring less than a billion trillion trillionth of an inch across. If such miniature wormholes really exist, then we might develop the technology to extract one, just as we can isolate single subatomic particles today. Then, conceivably, we could expand it to macroscopic dimensions, make it traversable, and finally maneuver its mouths into place so as to link any two designated points in space and time.

Of course, this description of how we might someday construct a network of wormholes for rapid transit across the universe is necessarily vague. Scientists are only beginning to scratch the theoretical surface of the problem. As for the technology involved, it simply lies beyond our present capacity to imagine. But then it would have been hard for anyone even a century ago to foresee the means by which robot probes would reach Uranus or Neptune and send back detailed pictures of those worlds.

Tunnels through space and time that bypass the usual

travel restrictions of the cosmos would allow us to colonize remote parts of the universe almost as easily as other planets and moons of the solar system. We could transport ourselves, and our thinking, self-aware brains, to other stars, galaxies, and galaxy clusters, and even to different periods of time. Yet, if this happens at all, it will be in the very far future—thousands, or perhaps even millions of years from now. By that time, as we shall see, it may not be our physical bodies that we choose to move around the cosmos, but merely our thoughts and sensations. In fact, it may be that a future wormhole network will be used not so much for transportation as for communication between the various parts of a collective, growing human mind.

There is one last and most extraordinary possibility that may bear on the future of life and intelligence. If we can become architects of space-time—if we can learn eventually how to make and harness our own traversable wormholes—then it may be that we can also create entirely new, viable universes. Essentially this would involve setting up the desired initial state of the new universe inside a wormhole, just as we can prepare, say, a primordial planetary atmosphere in laboratories today. We would be free to select whatever initial conditions we want. Then the ends of the wormhole would be pinched off to produce a completely separate region of space-time. The disconnected wormhole—now a new universe, or "babyverse," in its own right—would subsequently evolve from the initial state we prescribed. This evolution could involve a sudden, spontaneous outgrowth of the new space-time, similar to the Big Bang, which started our own universe, followed by a lengthy period of rapid expansion during which the babyverse might develop its own forms of matter and large-scale structures.

Having cut, as it were, the space-time umbilical joining it to the parent universe, we would free the infant cosmos to develop at a rate entirely outside our own time stream. It might, for instance, grow to be billions of light-years across in just a few minutes or hours as measured by our own clocks. To see the results of our handiwork we would have to reestablish a space-time link with the babyverse some time after its detachment. To do this, it would be necessary to keep track (by computation) of the whereabouts of the babyverse in hyperspace relative to our own cosmos. At a chosen moment, we would then create a wormhole bridge to the new universe in order to send instruments through to monitor its evolution and discover what long-term developments had ensued from our particular choice of starting conditions.

No doubt to begin with our babyverses would be stillborn. It probably takes a very carefully chosen (perhaps unique) set of genesis parameters to give rise to an "interesting" future cosmos—one in which stable particles form and eventually conspire to make objects such as stars and planets. We would learn, through our new science of experimental cosmology, how much—if at all—the familiar fundamental physical constants and natural laws of our own universe can be tampered with and still lead to a cosmos of rich and varied morphology. We could, in effect, test the various versions of the Anthropic Principle. In time, we might even discover how to prime a babyverse so that it went on to develop advanced life-forms. Through a subsequent space-time link, we could then establish communication with any intelligent inhabitants and let them know that we were their makers—in effect, their god. Perhaps, knowing how the trick was done, they would go on to spawn their own, third-generation universes. Or—a disturbing thought—perhaps all this has happened before and, in truth, our cosmos is someone else's babyverse. . . .

We stand at the threshold of a most astonishing future, whatever the details of it may turn out to be. In all likelihood, we shall push out into more and more distant reaches of space and time, bringing consciousness to regions that were previously unaware. But what form our future intelligence might take is open to question. It would certainly be nearsighted, given the vistas of time involved, to assume that man will not alter much in his physical appearance. In fact, our organic bodies and brains may become obsolete sooner rather than later. To imagine that the spread of man throughout the universe will always necessarily involve the translocation of our physical selves is naïve. In the long run, it may be the mere transference and communication of conscious thought across space and time that we hold to be important. If, in some sense, parts of our brains can be linked directly, intimately, to other centers of consciousness elsewhere—say, through a wormhole communications channel—then there is no need to make the journey "in person." The freeing of the mind from its organic shackles will be the most spectacular leap in the evolution of consciousness as a whole.

12

Man and Beyond

The mind of man is a communal affair, a product of many brains interacting with one another over many thousands of years. Seen in this context, the importance of language as an effective means of sharing and developing thoughts can hardly be overstated. We are each like a node in a network of billions of such nodes, many still active, others that have been active and contributed in the millennia gone by. Cut off from this network—this communal mind or consciousness—we would be almost as helpless as a solitary ant. But as part of the collective human mind, our intellectual potential is enormous. Nor is that potential limited to the spoken word.

Through books, the accumulated wisdom of our species has been put at the disposal of anyone who can read. In effect, the world's libraries serve as an annex—a huge, well-ordered, ever-growing annex—to human memory. As a result, what has already been discovered, invented, or imagined can be used as the basis for future forays in thought. Ideas can build hierarchically upon ideas. Moreover, books are not mere repositories of knowledge. In a genuine sense they are interfaces between minds. Their very familiarity and down-to-earthness hides their full significance: that they are a first step toward expanding consciousness further—through technology.

More recently, we have developed other ways to transfer information and ideas rapidly between human brains. The telephone and the various devices that exploit the global phone network (such as fax machines and teleconferencing systems) allow anyone with the right equipment, anywhere on the planet, to make almost instantaneous contact with anyone else. A thought that occurs in your brain at this moment can be communicated to another brain in Moscow, or London, or Jakarta within seconds.

Likewise, we now have the means to relay information—both verbal and visual—to hundreds of millions of brains at once from a central broadcasting point. Through television, radio, and satellite links we can become aware, not just as individuals, but as nations, of events taking place live at any point on the Earth, or beyond. As a consequence, not only are we better and more quickly informed (or misinformed) individually, but also our consciousness as a species is raised. Man's "superbrain," its neurons comprised of individual human brains, is becoming much more coherent and efficiently interconnected because of electronic mass media.

Of course, collective consciousness in some form has existed on Earth for almost as long as consciousness itself. The apparent corporate sentience of social insects, such as ants, termites, and bees, is well known. Certainly, our hominid ancestors depended for their survival on the knowledge and skills contained within their group mind. Yet it is only very recently that this type of awareness has become integrated at all on a planetwide scale.

The collective human psyche is in its infancy, unsure of itself, and, like all newborns, still largely uncomprehending of the strange new sights and sounds flooding into its mind. Not surprisingly, considering its immaturity, it responds erratically and often inappropriately to the information reaching its senses—of revolutions, of human and ecological disasters, of other happenings on the world stage. There are inconsistencies, for example, in our corporate approach to issues like global warming and economic growth. Individually, we may be capable of thinking such problems through and arriving at solutions in a coherent way. But our collective mentality is so fresh and undeveloped that it naturally seems childlike by comparison.

Yet, as it matures, our communal consciousness can be expected to become vastly more powerful and important than our private minds. The step up from individual consciousness to genuine planetary awareness is analogous to the difference between a single neuron and a whole brain. Just as one neuron cannot conceive what it would be like to be a collaborating brainful of neurons, so, on a personal level, we cannot begin to grasp how a planetwide consciousness might feel. Each neuron plays its part in a thinking brain, without itself having sophisticated thoughts or knowing anything of the experiences of the whole brain. In the same way, each of us is becoming part of a larger mentality without personally being able to imagine what it will be like to be that entity. The only way we can experience collective consciousness is to merge with it and—at least, temporarily—give priority to it over our individual selves. In a limited sense, that is what we do whenever we interact with other people, whether it is face to face or by telephone or, passively, through television or radio. During such an interaction, our awareness of self diminishes and we become instead part of a broader mental community.

This tendency to participate in a communal consciousness—to simultaneously share thoughts with others—is likely to continue to grow rapidly over the coming decades. And it will be encouraged not just by the availability of books, newspapers, television, and an expanding, improving telecommunications network. These are vital aids to information transfer. But they do not, in themselves, add to the collective processing potential of the human race. To help man greatly increase his communal brainpower a completely different type of device is needed: one that has the capacity to accept and carry out instructions on its own, that has a memory for storing coded tasks and data, and that has, like a human brain, the talent to distill new information from old.

We call it a computer, but that is a poor name for an invention of almost limitless power. True, the first computers did merely compute. They added numbers and juggled statistics a thousand times faster than any human tallier could. But even before the end of the Second World War it was clear that computers were capable of a great deal more. In theory, they could emulate the many ways in which the brain itself can process information—and, in time, far exceed the brain's potential. They could be faster, more retentive, more creative and, eventually, perhaps more intelligent than ourselves.

So began the first attempts to create a machine mind, though this was a technological quest of an entirely different order from that to land astronauts on the moon or to unravel the physics of black holes. Its goal: nothing less than the total synthesis of intelligence, an artificial power to rival or surpass that of the most intricate known natural system in the universe.

Four billion years it has taken for man's neocortex to evolve by unconscious selection. But now a new, alien intelligence has begun to spread across the face of the earth. This is an intelligence born not of slow biological evolution, but of man's own, more sudden handiwork. For the communal human mind today is engaged in consciously, purposefully devising an inorganic extension of itself—and it is making spectacular progress. In processing speed, memory capacity, and component density, computers have advanced exponentially over the last fifty years. They have become weather forecasters, fusion simulators, Bohemian artists and music makers, molecular microscopes and designers extraordinaire. Nourished only by streams of cryptic commands and data they can defeat human grandmasters at chess and interpret visual scenes captured by robot eyes, almost as if they are already aware, if not of what they are, then at least of what they do.

And yet, true artificial intelligence does not seem to be just around the corner. It is one thing to fashion an information processor of great speed, and quite another to imbue it with perception, imagination, creativity, awareness, and all of the other remarkable, elusive higher talents of the human brain. In fact, it is becoming clear that artificial intelligence may never be achieved overtly. That is to say, it may be unrealistic to hope to cast transistors, or Josephson junctions, or any other kind of switch, into some array, and then rigidly program that array to behave like a human brain. It might take centuries for the brain to map itself completely at the cellular level, to unravel the chemistry of thought and the physics of consciousness—and only then could it begin to draw up blueprints for an inorganic protégé of itself.

Existing computers are too inflexible and simple in design to meet this new challenge. Serial assemblies of primitive, two-state switches, they must be instructed laboriously, step by step, to reach any goal. Yes, they can carry out repetitive calculations brilliantly. And yes, if we knew how, we could perhaps program them to do the very things at which our own brains excel—but it might take a thousand years. In any case, there is a far easier way.

Imagine a complex, three-dimensional network of electronic processors, laid out like the neurons and synapses of the brain. Each processor, or node, has a large number of input paths, among which it can distinguish, and a similar number of output paths. Each node also has a program by which it can vary the pattern of signals it sends out according to the pattern it receives. And lastly, each node has a means of changing its program and of receiving a signal that may say effectively: "Good, keep the program as it is" or "Fair, modify the program slightly" or "Poor, alter the program radically."

Now suppose that this network is exposed to many

different signal patterns, or sets of raw data. To each the system responds, producing an output pattern. If it responds well—if it answers correctly or nearly correctly—it is praised, so that its existing program is reinforced. Otherwise, it receives a reprimand and subsequently revises its programming.

From being totally inept, the network progressively learns from its mistakes until it can perform with uncanny accuracy. There is no external programming or preordained solution method, only a process of learning, as a child might learn, from experience. The human "educator" may not know, or even have any way of finding out, just what the connections are that permit the array to recognize a pattern and to respond appropriately. That educator may not know what the current program is in any given node. The network develops internally, unseen, and of its own accord.

Remarkably, such networks, known as neural nets or neurocomputers, have already reached an embryonic stage. Present research is focused primarily on stimulating their behavior, using very powerful conventional computers. But the first crude, three-dimensional arrays of artificial neurons have been constructed—and the results from them are indeed promising.

At Warwick University, in England, for example, researchers have developed a neural net for recognizing odors that works in a way similar to that of the human nose. Its biological equivalent—a poorly understood organ—has between six and twenty varieties of smell receptors with about a million copies of each; our brains identify smells from the pattern of responses these receptors generate. The Warwick electronic "nose" employs twelve receptors, in which the electrical resistance of a film of tin oxide varies with the concentration of gases surrounding it. The output from each sensor enters one neuron in the first layer of a triple-layered neural network. Each of these neurons is linked to all the neurons in the concealed middle layer, which, in turn, are joined to all the neurons in the third, output layer. Neurons, whether electronic or biological, multiply the signal on each of their inputs by a weighting factor and then pass the sum of the results to the next layer. The researchers at Warwick trained the artificial nose to recognize smells by presenting it with five different alcohols, and adjusting the weighting on each link in the network until it produced a consistent pattern of outputs for each substance. Eventually, the device could find a number of practical uses, including monitoring food freshness and sniffing out waste products in the environment.

Scientists from IBM and Cornell University, meanwhile, have developed a software model of ten thousand brain cells, which spontaneously produces signals similar to those from the brains of resting animals. Their current design imitates a slice of part of the brain called the *hippocampus,* which is essential in the function of memory and is also involved in the development of disorders such as epilepsy. The connections within the simulated network are based upon experiments in which drugs were used to block selected routes between the cells of real brain tissue. The team tested how often the signals got through those connections that were still open, and worked out a statistical value for the most likely number of signals each cell must have received. Neurobiologists have known for some time that a certain class of brain cells—pyramidal cells—fire spontaneously even when at rest. The resulting signals these cells produce are closely matched by the resting signals put out by the IBM-Cornell model. No one really understands the cause of these signals, known as population oscillations, either in the model or in the brain itself. Although the brain contains billions of neurons, each is attached to only a very small proportion of those around it. The difficulty is to explain why the cells appear to act in unison when the connections between them are so sparse. Future research at IBM and Cornell may shed light on the origin of population oscillations and, in general, the way large collections of brain cells function *en masse*.

Little by little, computer scientists hope to make neural nets behave increasingly like organic brains. As a result, in those areas at which the human brain is especially adept—in making sense of complex visual scenes, in comprehending natural langauge, in solving abstract problems, in all forms of concurrent and creative processing—neural nets, too, may come to excel. These are precisely the areas that we hold to be most significant in gauging intelligence. Of course, speed of calculation is also important for many purposes, and for these applications a powerful, serial computer running orthodox programs may remain the obvious solution. But neural nets offer the possibility of genuine artificial intelligence. Moreover, they are in no way limited, as an organic brain is, by the number of nodes or interconnections they can accommodate. As the scale and sophistication of neural nets progresses, so will their intelligence, until finally they may exceed the intellect of their creators.

In all likelihood, long before that happens, neural nets will be enlisted to help design and educate their successors. Within half a century there may be desktop versions of such machines that we consider to be as much our mental partners as our tools. Imperceptibly, our children and our children's children will come to regard computers in a new and very different light, not as boxes of wires and transistors that are only as clever as the person who programmed them, nor as swift but stupid slaves. They will see them, instead, as true mind machines—and eventually, simply, as companion minds.

This is only one aspect of the impending revolution. Already other developments are in motion that, within a generation or two or three, are likely to lead to a broadening and amplification of consciousness on this planet. Soon we may be privy to new, undreamt-of realms of experience as our individual minds, and the communal mind of man, merge with the thinking devices we have built.

Even now, equipped with special, sensory goggles and gloves, one can become a traveler in the strange land of "virtual reality." An outgrowth of interactive models of environments, such as flight simulators, virtual reality allows the user to enter and experience the computer's inner world. Sensors in the goggles detect which way the traveler is looking while other sensors in the gloves monitor finger positions. Any physical movements are thus instantly translated into corresponding movements in the computer's fabricated universe—a universe without frontiers.

Thanks to virtual reality, an architect could stroll down the passageways of a building yet to be constructed. A medical student could swim along a simulated vein to a simulated heart, ride a nerve impulse to the brain, or rehearse an operation without fear of injuring a real patient.

"Trippers" might succumb to the perils or pleasures of extended sojourns through Tolkienesque Middle-earths, assuming a different heroic role each time.

And that is likely to be but the start of our close association with computers. Researchers are already looking at ways of replacing today's clumsy electronic goggles and gloves with less obtrusive interfaces. As a result, it may not seem so unnatural in the future to link our organic minds intimately and for increasingly long periods to the minds of our invention. For instance, we may eventually be able to communicate directly with neural nets anywhere on Earth via microscopic transreceivers implanted just below the skin. These transceivers, in turn, might connect to the brain's primary data highways—the nerve bundles of the corpus callosum that join the right and left hemispheres. More and more, it seems, we shall view computers not as elaborate adding machines but as fellow minds—and perhaps even as superior minds—from which a close association can only bring immense reward. Through direct brain-computer links we shall have access to all the speed, the penetrating intelligence, the knowledge, and the virtual-reality-creating capability of any or every computing device in the world.

Since, if this happens, every individual could be in constant and intimate communication with the global computer network, each human brain would also have the means to contact directly every other—to literally read another person's mind, to actually meld with that mind. And just as passwords are required to access private computer accounts today, so a similar scheme might operate in the future to allow people to choose between personal contemplation, more extended mental discourse involving a select number of other minds, or global thought sharing involving the whole planet.

A communal system interfacing human brains with powerful computers would open up some staggering possibilities. Every observing and sensing instrument at man's disposal could be interfaced to the emerging planetary mind: telescopes offering for immediate sensation the most delicate, star-studded filigree in the arm of some remote spiral galaxy; particle accelerators furnishing esoteric data from which neural nets could synthesize virtual realities of the observed quantum world; robot space probes beaming back images of virgin star systems, their camera-eyes effectively our eyes; X-ray, infrared, and ultraviolet detectors feeding in images of the cosmos beyond the visible. These would be our new and unimaginably powerful senses, channeling data directly to mankind's corporate mentality, as today our five biological sense organs supply our personal brain.

Looking forward a few centuries or millennia from now, there would still presumably be people, and there is no obvious reason why they should look any different from us. Yet they would each be part of an astounding, evolving consciousness. They would be telepaths by any other name, mental Leviathans, atoms of a common, holistic mind, still growing, still exploring its vastly enlarged intellect and awareness.

Today, that prospect may seem unnerving and totally alien. But to those who grow up in the new world, it will be we who are pitied for our impoverished, solitary lives. Future generations may look back and wonder how we could bear to be so horribly alone, prisoners of our skulls and stunted senses. To see from radio wave to gamma ray, to walk with friends and share their innermost thoughts in lands real and imagined—these are just some of the experiences that we can hardly begin to appreciate.

A close coupling between brain and computer also leaves open the possibility that at least part of a person's consciousness could survive indefinitely. As an individual's organic brain faded and failed, its memories and personality could perhaps be preserved in imperishable form—not as some nightmarish tangle of metal and wire, but as a living intelligence within the expanding community of minds.

From our standpoint, in the twentieth century, we can only guess at the directions this blossoming superhuman intellect might take. But there would inevitably be enormous freedom of choice. Each component of the collective mind, each "person," could blend with the holistic consciousness or assert its individuality at will. As the organic receptacle—the body—of the person wore out, it could be replaced with an identical, genetically engineered replica. Or its substitute might be a body of a different design, or a nonhuman form—perhaps an interstellar probe, so that one mind, or a family, could travel to other stars, even to other galaxies.

And as mankind did spread throughout space, both physically and mentally, so planetary minds would spring up around countless alien suns. Each mind would be a community unto itself, but would also be in intimate contact with other planetary minds, just as the cells in a human brain continually commune with each other.

To begin with, thoughts between worlds might be exchanged along conventional channels, that is, by signals traveling at the speed of light through ordinary space. But, in time, such pedestrian means will surely be superseded if it is at all possible. A far more effective communications network could be based upon wormholes, providing that these tunnels through space-time can be adapted or built to our specifications. Relay stations sited at each end of a wormhole would allow planetary minds to exchange thoughts with virtually no delay. Even a consciousness located around a star on the other side of the Galaxy could perhaps be contacted within seconds.

Given that a telecommunications system using wormholes might eventually become a reality, there would be no need for intelligence to remain confined within our own galaxy. Distance would become no object. If signals are able to travel outside of ordinary space and time, then they take no longer to travel a billion light-years than they do to go a mile. So we can extend our vision of the future to include consciousness spread not just all around our home Galaxy but around the entire universe. In fact, we can begin to glimpse a complete hierarchy of consciousness evolving out of this: billions of neurons making up a human mind; billions of minds on Earth contributing to a planetary mind; billions of planetary minds comprising a galactic mind; and billions of galactic minds resulting, eventually, in who-knows-how-many-eons-from-now, a mind cosmic in scope.

As all this happens—and it is, even now, starting to happen—we shall come to know the true profundity of nature. Our present experiences and sensations are likely to seem extraordinarily shallow as we evolve X-ray eyes and organs so sensitive that, for example, they can feel each passing gravitational wave. And this heightened consciousness will allow us not only to experience more and more of the external universe, but also to create and explore new worlds of our burgeoning imagination.

We shall begin to appreciate and to sense firsthand the multilayered nature of reality. We shall come to understand that consciousness and intelligence derive, at root, from a mathematical play of matter, a complex, nonlinear dance of the raw stuff of the world. It will become apparent that when a mind thinks, it alters its own pattern, and thus not only becomes effectively a new mind but changes the very state of the universe. That coupling may not be significant as long as the amount of matter incorporated in mind is small. But it may become much more important as intelligence propagates throughout the cosmos until, in the farthest future, it may become the dominant universal influence. Then, as the cosmic mind thinks, its substance—which ultimately may include all the substance of the universe—will spontaneously reorganize itself. Our dreams and ideas and abstractions will manifest themselves immediately in reality.

In the much shorter term, there are possibilities little less startling. If all that we personally are can be encapsulated and stored and replicated at will, then there is nothing sacrosanct about humanoid intelligence. We are simply intelligence, without qualification. And we can create other intelligences that, from the outset, are divorced from any specific hardware, organic or otherwise. We can create wholly numerical, thinking beings from scratch—because, stripped of flesh and bone and organic brain, we are simply that ourselves—numerical beings, whirlpools of self-perpetuating complexity, collections of interacting mental models that can thrive upon any suitable substrate.

Imagine a time when working together within our planetary supermind we can invent new intelligences—pure numerical intelligences inhabiting a numerical universe that is a subset of the global mind. Would these mathematical occupants of a mathematical cosmos, exploring in their mathematical spaceships, be real or just models of our imagination? And how would their perceived reality differ from ours?

There is no reason why an individual who was spawned as an artificial intelligence should not, just as we could, employ an organic brain and body to emerge into the outside universe. What previously would have been an invisible meta-cosmos to that creature would now be revealed and experienced as the familiar space we know today.

The truth is, reality in the future will seem immeasurably richer and more complex than it does today. We shall be no different, eventually, from the intelligences we manufacture, so that we shall have the same freedom as they. We shall be able to pilot our consciousness physically around the old, human universe of stars and galaxies and sandy beaches if we wish to. Or we may choose to dive into the underlying mathematical world in which "virtual" objects and events can simply be imagined into existence, or to plunge further into an endless series of other mathematical strata. And just as we shall have the capacity to create artificial life, so this life, within its own simulated cosmos, would be able to create "virtual" artificial beings, and so on, without end. And each of these beings would have the same liberty as ourselves—we the ancient ones from the original universe of flesh and blood—to travel up and down the multifarious levels of reality.

* * *

Such prospects are dizzying, and we can hardly expect to grasp yet all the implications of being part of some future communal mind. We are perhaps very much farther removed from such a state than we are from, say, the mental prowess of a fly.

But this much is clear. The potential exists for mind to become vastly more extensive and capable in the future than it is now. Even in the short term, it seems that consciousness is set irretrievably to spread beyond the Earth and beyond the organic circuitry of a few billion bipeds. Over much longer periods, mind may grow and integrate over much of space and time, so that effectively the universe as a whole becomes conscious. Much depends on how far the laws of nature will allow our technology to evolve. But if the universe can become globally aware—if a large portion of all the matter that exists can eventually participate in intelligent thought and observation—then it is hard to see how the powers of mind can be overestimated.

13

The Watcher at the End of Time

We look out today and see a universe the meaning and materiality of which seems contingent upon our own existence. "What is man," asks John Wheeler, "that the universe should be mindful of him?" The more we learn, the harder it is to avoid the extraordinary conclusion that nature has in some sense been waiting for us.

We inhabit a very special type of cosmos—special not just because it supports life, but because if the laws of nature, the values of the basic physical constants, and the initial cosmic state had been even marginally different, then the evolution of life of any kind would have been precluded.

We live, too, in a universe which, if we accept the standard Copenhagen interpretation of quantum mechanics, allows us to play a major hand in determining its contents. Only during the process of observation, in this view, does the wave function of an object collapse, its "hereness" and "nowness" crystallize. Furthermore, this applies not just to subatomic objects, such as electrons, that we might try to disregard. It applies to everything that has a wave function: to trees, bottles of Burgundy, stars, galaxies, clusters of galaxies, and, inevitably it seems, the universe as a whole. When we make a measurement, when we observe any facet of reality, we participate in establishing what that reality is.

As we have seen, there are other, less anthropocentric viewpoints, such as the Many Worlds interpretation of Everett. But if we are willing at least to entertain the notion that consciousness may be a significant factor in shaping the cosmos, then we can glimpse some extraordinary possibilities. Indeed, we can begin to sketch out a remarkably comprehensive and self-sufficient model of how reality may work.

Mind seems so utterly trivial. After all, there are little more than five billion of us humans, with a total brain mass not exceeding fifteen billion pounds. The earth alone weighs a thousand trillion times as much, the solar system a third of a million times as much again, and the universe ten billion trillion times as much yet again. Materially, that part of the universe engaged in high mental activity is vanishingly small. But in considering the effect that mind may have on cosmic affairs, the combined mass or size of our brains today is totally irrelevant. What is crucial is the brain's complexity, the extent of its interconnectedness, and its consequent capacity for rational thought. Gauged in these terms, the human cortex outstrips every other structure in the known universe. It can do what not even a galaxy, or a cosmos of galaxies, alone can do.

We need to remember also that one of the brain's tricks is to conceive of itself as an individual, as something apart from the rest of the universe. That was an inevitable outcome of its acquired survival skill for labeling and analyzing the world around it. The brain habitually thinks of itself as a discrete unit. But to grasp the true situation, we need to switch to holistic vision. We need to conceive of our brains, not in isolation, not as belonging uniquely to "us," but as an inseparable aspect of the universe—for that is what they are. Ten billion years ago, the atoms now temporarily patterned as your body and mine were strewn across many cubic light-years of the Galaxy. By chance they came together in an interstellar cloud and were then further condensed into the material of the Earth, debris left over from the Sun's formation. For billions more years they combined and split apart and recombined in multifarious ways with other atoms in the Earth's crust, some of them occasionally drifting in and out of the anatomies of other animals and plants. Almost certainly, a few of the atoms now helping you interpret these words were, for a brief spell, incorporated in some dinosaur or Carboniferous fern and, well before that, the deep interior of a monstrous star. Each particle of your substance has a unique and fabulous tale to tell.

Nor is the eons-old matter from which you are made unchanging. With every breath you take, you gain and lose so many trillion atoms. You are not the same person you were a year ago, or even a millisecond ago. You are not a thing but a smoothly flowing process. Things—changeless, isolated entities—are only fictions of the rational mind. We have evolved the skill to conjure up such distinctions, to invent objects and names, in order to make sense of the world. But we need to remember that these "things," including you and me, are indeed mental constructs. What is a person? Certainly not a particular collection of atoms, in some special, frozen arrangement, because from one instant to the next that collection and arrangement changes. And manifestly, too, a person is not detached from the outside world, for we interact continuously with our surroundings. A torrent of photons from the Sun and every other object around you is absorbed and reemitted by your body each moment, so that you are linked intimately, continuously, to the source of all these particles at the quantum level. To this extent, you are irrevocably tied to the Sun, the Earth, and the stars. You are an integral part of everything—the whole cosmos—because the cosmos is physically indivisible.

When you think, the cosmos thinks—not in some nebulous, poetical sense, but literally. Your brain is a product and a process of the cosmos. When you look out and form pictures in your mind and try to make sense of those pictures, you are the universe trying to make sense of itself. Forget that we sometimes imagine ourselves to be tiny bipeds on a ball of rock somewhere in the awful depths of space-time doing cosmically trivial things such as changing a diaper, or mowing a patch of grass, or brewing the next cup of coffee. The incredible and undeniable fact is that we are the thinking, reasoning components of the universe—a realization that makes it slightly less astonishing that our brains might have some role to play at the cosmic level.

We need to bear in mind, too, that we almost certainly

represent just the first sparks of awareness. Our brains may be the most ornate and sophisticated fragments of matter ever wrought (as far as we know). But we have only been remotely human for a few million years, for less than a tenth of one percent of the age of the cosmos. We have had a substantial language for perhaps only the last forty thousand years, an organized civilization for under ten thousand years, and a comparatively advanced technology for a mere century or so. To what far loftier heights might our intellect—the intellect of the known universe—soar over the next million years? Or the next billion?

We are about to witness, and be part of, an explosion of mind. And yet, paradoxically, as a species, we have also come to the end of our biological evolution. A thousand centuries or more ago our brain ceased to evolve, for it was then that the first stable, reasonably secure human communities began to appear. With members of a tribe pooling their skills for the good of all, individuals—whatever their strengths or weaknesses—were shielded from evolutionary pressure. Safe within your commune you are as likely to breed and so perpetuate your genes whether you are a fool or a genius: a situation that remains unchanged to this day.

Before it reached its contemporary plateau, however, the human brain achieved a certain level of complication, organization, and size. Language, rational thought, and unbounded creativity became possible, even habitual, so that we humans gained the means by which we could evolve further, not biologically this time but technologically.

With their newly acquired talent for analyzing and labeling the world around them, the first modern *Homo sapiens* began that frenetic phase of postbiological development that continues even today. They tamed fire and fashioned tools of stone, bone, bronze, and iron. They slew at will what wild creatures they needed for food and clothing, and later domesticated the more useful of animals and plants. They painted and sculpted and propagated their traditions in ever more refined speech. In time, they set down their thoughts and knowledge as symbols on stone and papyrus, so that the sum total of human wisdom was no longer constrained by what could be held at any moment in living brains. Knowledge, plans, theories could be stored extrasomatically for future use. And there were many ways of writing—those of the poet and playwright, of the historian and scientist, and most arcane, but profound, that of the mathematician.

No more would man's old organic brain expand. But it had no need to, for it already held all the potential to broaden human consciousness without itself changing. The continued intellectual evolution of man, the spread of knowledge and understanding and awareness, could now be achieved by other means, by the conscious manipulation of nature. Man took the world apart, both with his hands and with his mind, like a child disassembling a clock. He analyzed relentlessly, learning more and more of the rules by which nature played its game. And he was astonished to find that these rules were mathematical. The universe seemed to be a colossal enactment of a drama written in the language of differential equations, a performance in which man, and the substance of which he was made, was intimately involved.

Analysis was only the first step, the first half of the loop. Having broken down the material world, man could then reassemble the pieces in new ways; so, as he evolved, he made portions of his surroundings evolve, too, at a vastly accelerated rate. He fashioned new materials and machines, including instruments with which he probed nature still more deeply to reveal aspects that lay beyond the reach of his biological senses.

Through telescope-enhanced eyes, working in the visible spectrum and beyond, man became aware of the vastness and antiquity of the cosmos. But he became aware, also, that it seemed to have had a definite beginning: an explosive moment of origin in the remote past.

That was the first clue to the mystery of how and why the universe had come about. But another clue was to be gleaned by looking in a very different direction: not outward to the farthest limits of the macroscopic, but inward, ever inward, to the inconceivably small realm of the subatomic. In fact, it was from here that the greater revelation was to come. In broad concept, the Big Bang had been anticipated—so well did it accord with both a rational secular view of creation and a liberal reading of the Biblical genesis. But by contrast, the message coming from the quantum world, as interpreted by Bohr, was unexpected, bizarre. In essence, it was that reality without observation was also without meaning; that there is no such thing as material particles without the conscious involvement of an observer.

Suddenly, man found himself bewildered, on center stage. One moment he had been a tiny, self-effacing figure in the audience marveling as the epic play of the universe unfolded; the next, he was blinking into the footlights with the universe sat before him—the star of the show.

We are not, the Copenhagen interpretation of quantum

mechanics implies, a pawn or a trivial spectator to a cosmos that owes nothing to our existence. On the contrary, it is apparent that nature depends quite crucially upon our participation. And it does no good to argue that these subatomic systems we are supposed to help create are but little things, able to be ignored if we choose. The fact is, everything in the universe, including our own bodies and brains, is composed of elementary particles. If conscious observation is the key to reality making on the smallest scale, then inescapably it is crucial, too, on the cosmic scale. In the final reckoning, everything there is—the whole universe—owes its existence to countless individual acts of observation. Seen in this light, it is hardly surprising that the cosmos appears uncannily well suited to the evolution of intelligent life. *We had to be able to evolve in order that the universe could exist.*

Yet there seems to be an obvious flaw in this argument. The problem centers on those times when there were no "reality making" human beings, when there was no mind, or even life, of any kind. How could the things that presumably existed then have been made real so that we could eventually evolve from them?

To understand the way around this difficulty, we have to adopt a radically new picture of nature. We have to see that the universe may exist as a block, throughout the whole of time—and that it has always done so. It never actually did begin. In other words, there never was a Big Bang in the conventional sense, cut off from everything that would follow. All the universe, past, present, and future, exists at once, as a closed, self-sustaining, self-creating cycle. It never started, and it will never end. It simply is.

In fact, there is nothing unnatural about such a concept—on the contrary. Babies do not suddenly or inexplicably appear from nothingness. They come from parents who were once babies themselves. Water does not spring out of some timeless void to fill streams and rivers. It evaporates from the sea, forms clouds and falls as rain on the mountains. Rivers come from clouds that were once rivers themselves. And it is the same with the air we breathe, with the stars, with the very atoms of our bodies—all are self-recycling. So we should not be so taken aback if such a process of self-regeneration were found to apply to reality as a whole.

Once we become comfortable with the idea of the universe as a colossal feedback loop, then it is the notion of a unique creation event that starts to seems strange. It is really the traditional belief that the universe was sparked off at some particular, initial moment—either naturally or divinely—that introduces all the most perplexing cosmological problems. What was there "before" the universe, how did the first event manage to occur outside of time, how was the specific and remarkably well-tuned nature of reality chosen? Such difficulties vanish as soon as we allow that genesis, conceived of as a primordial event that happened long ago and is now over with, is an illusion.

Only in our minds is the world made real. And, simultaneously, only in our minds does the mathematical infrastructure of what we see take form. We look out and see an environment—a cosmic environment—that must be such as to be capable of nurturing us. Mind selects, defines, and refines the nature of the reality which must inevitably lead to mind. We are the universe in dialogue with itself: our doubts and discoveries, our truths, small and large, are all forms of the cosmic drive toward clarity and truth. Through us, the universe questions itself (as in Wheeler's parlor game) and with each response edges a little closer to the physical being it must become in order that it can exist.

We stand at the merest beginning of that stupendous process. Our minds, still feeble and small, are presently capable of only a tiny amount of creation through participation—of wave-function collapse. Our influence is only just beginning to be felt as our analyzing, symbolizing, comprehending brains uncover the patterns of nature and as our technology gives us the power to penetrate more deeply into the fabric of things. With the growth in scale and sophistication of that technology, human consciousness will extend to levels that today we find impossible to imagine. Future computers, in particular, may form the basis for a communal mind that will spread, first over this planet and then beyond, seeding ever more remote worlds with an intelligence whose scope is beyond our ability to grasp.

Of course, it seems far-fetched. But then the very existence of the universe is far-fetched—however we try to explain it. The abilities of the human brain are far-fetched. (How can a three-pound lump of flesh imagine and see the world as it does?) The fact that we are beginning to make sense of a universe so incredibly vast and complex is far-fetched. There is no answer that you could conceivably give to the deepest mysteries of all that is not so outrageously fantastic as to be almost beyond belief.

Today, our minds are still immature, our consciousness still personal, "atomic." We each think of ourselves as individuals riding the wave crest of now that separates the future from the past. We consider the past to be over, and the future as yet nonexistent—uncertain, unknowable. And, indeed, that is the exact truth of the matter as far as we are concerned. We do have free will. We can affect our personal and corporate tomorrows. Only a creature that stood outside the universe—a supernatural god—could see it otherwise. But there is no necessity for such a being. The universe can act as its own god, supplying the means of its own creation, without external help. We are a part of that all-encompassing entity, though as yet only a trivial and embryonic part. The fact is, if we are just now beginning to flex our creative powers, then there must be other, much more powerful influences helping to make us, the stars, the galaxies, and the Big Bang itself real. These influences can only be a result of the greater communal mind of the future.

As mind grows and its power to observe and comprehend nature at all levels expands, so will its ability to watch quantum systems in the past. Thus, from increasingly remote points in the future, an increasingly powerful creative beam of observation will reach back to earlier and earlier times. And so will the cosmic mind-to-come project more and more into the past the very reality that is necessary and sufficient to give rise to it. As the Final Anthropic Principle suggests, this will go on until, in the infinitely remote future, there is a single all-encompassing mind, coextensive and coinclusive of the entire universe, that through a single, final observation completes the loop joining alpha to omega: the watcher at the end of time.

We may not have answered all the mysteries of the universe by means of this bold new teleology. That would be too much to expect. But we have at least provided an all-inclusive framework for linking such puzzling concepts as matter, mind, mathematics, and the "meaning" of reality. The picture that emerges is of a completely self-contained loop in which consciousness—both now and in the future—plays a pivotal role in actualizing the cosmos, and therefore itself. Our inner world of consciousness is effectively the sieve through which all that is possible must pass in order for the one physical reality out there to be selected. The mathematics of that reality is the one logical system that, taken as a whole, can underpin a universe in which conscious beings can effect such a selection. Matter is the substance of that world—the stuff by which the logic and meaning of the cosmos is made manifest to our senses. But, at its most fundamental level, matter is merely mathematical. Things are mathematical structures, nothing more. Quantum particles, the basic building blocks of the universe, are mere probability functions. Only at the moment of observation are they localized, made "real." And only in the conscious mind of the observer are they bestowed with substance. Everything—every aspect of mind, matter, and mathematics—exists at the courtesy of everything else.

And man? It seems that we may be the very reason and the purpose that there is a universe—we, our progeny, and any fellow intelligent races that have sprung up elsewhere in space. If so, then we can look to the future with confidence and optimism. Even though our personal consciousness may dissolve at the point when our brains die, we shall inevitably be involved in the cosmic consciousness that is to come. Everything that we ever were, throughout our human lives, will gradually be reincorporated into the spreading awareness of the universe. Every particle that is part of you now, or was ever part of you, will eventually be reconstituted within this extraordinary, growing cosmic mind. We shall all live again, and be very much more alive and conscious than we are today. Nor will that greater consciousness ever end; on the contrary, it will change and evolve and expand throughout all space and time.

What will it be like to be that future cosmic mind—to be God by any other name? We shall simply have to wait and see. But it will mean the final unification and reconciliation of our two ways of seeing the world: the scientific and the mystical, the left brain and the right, the male and the female. Because when mind includes every particle in the universe—when mind is everything—whatever is observed becomes instantly and physically a part of consciousness. No longer is there any barrier between inner and outer, between the experience of oneness with reality and the rational comprehension of its parts. The circle will be closed, the universe complete; and the equations of eternity solved for all time.